BOUNDS AND CHARACTERIZATIONS OF RECORD STATISTICS

BOUNDS AND CHARACTERIZATIONS OF RECORD STATISTICS

MOHAMMAD AHSANULLAH
AND
MOHAMMAD Z. RAQAB

Nova Science Publishers, Inc.
New York

For permission to use material from this book please contact us:
Telephone 631-231-7269; Fax 631-231-8175
Web Site: http://www.novapublishers.com

LIBRARY OF CONGRESS CATALOGING-IN-PUBLICATION DATA

Ahsanullah, M. (Mohammad)
Bounds and characterizations of record statistics / Mohammad Ahsanullah, Mohammad Z. Raqab.
p. cm.
Includes bibliographical references and index.
ISBN 1-59454-538-3
1. Distribution (Probability theory) 2. Conditional expectations (Mathematics) 3. Boundary value problems. I. Raqab, Mohammad Z. II. Title.
QA273.6.A43 2005
519.2'4--dc22 2005017124

Published by Nova Science Publishers, Inc. ✣ New York

CONTENTS

PREFACE

This manuscript attempts to present in an organized manner characterizations and bounds of Record Values. There are several interesting articles presented in various journals and in some books on characterizations and bounds of Record Values. Our aim is to provide a through presentation of these topics in this book. The book will be helpful to any one considering an independent study for characterizations and bounds of record values. This book presents the essential topics on the bounds for expectations of records and characterizations of probability distributions via record statistics.

The book is written on an intermediate level which needs the knowledge of probability theory, mathematical statistics and mathematical analysis. Knowledge of reliability theory would be useful. A brief introduction to record values is presented in the first chapter. Despite the connection between the bounds and characterizations problems, Chapter 2 through 5 can be studied independently. In these chapters, optimal bounds based on projecting some functions onto convex cones of non-decreasing functions are presented. Significant works on bounds of records by Danielak, Danielak and Raqab, Gajek and Okolewski, Raqab, Raqab and Rychlik and other researchers were discussed. Chapters 6, 7 and 8 deal with the characterizations of distributions using records of continuous as well as discrete distributions. The topics on bounds or characterizations covered in this book will be useful for readers and researchers in that direction. The researcher can figure out many open research problems while reading the results in the book. The readers interested in more detailed theory of records can consult Ahsanullah (2004), Arnold et al (1998) and Nevzorov (2000).

Summer research grant and sabbatical leave from Rider University enabled the first author to complete part of this book. The first author thanks Janet Miller, Barbara Damico and Joanne Wetherbee for help in the process of typing the manuscript. M. Raqab thanks the University of Jordan for supporting his research works. He also expresses his sincere thanks to his wife Sumayya for the longstanding support and bearing with him during the preparation of the book. We are grateful to Nova Science Publishers for willing to publish this manuscript.

September 2004
Mohammad Ahsanullah
Mohammad Z. Raqab

Chapter 1

RECORD VALUES

We introduce here some basic properties of record values that will be helpful for the bounds of the moments and the characterizations of distributions based on record values.

1.1. DEFINITION OF RECORD VALUES

The record values model was introduced by Chandler (1952). Suppose that $X_1, X_2, \ldots$ is a sequence of independent and identically distributed random variables with cumulative distribution function (cdf) F(x). Let $Y_n = \max(\min)\{X_1, X_2, \ldots, X_n\}$ for $n \geq 1$. We say X_j is an upper(lower) record value of $\{X_n, n \geq 1\}$, if $Y_j > (<) Y_{j-1}, j > 1$. By definition X_1 is an upper as well as a lower record value. One can transform the upper records to lower records by replacing the original sequence of $\{X_j\}$ by $\{-X_j, j \geq 1\}$ or (if $P(X_i > 0) = 1$ for all i) by $\{1/X_i, i \geq 1\}$. The lower record values of this sequence will correspond to the upper record values of the original sequence.

The indices at which the upper record values occur are given by the upper record times T_n $n \geq 0$, where $T_n = \min\{j | j > T_{n-1}, X_j > X_{T_{n-1}}, n > 0\}$ and $T_0 = 1$. The record value sequence $\{R_n\}$ is defined by $R_n = X_{T_n}$, $n = 0,1,2,\ldots$, with $R_0 = X_1$. The upper record times of the sequence $\{X_n, n \geq 1\}$ are the same as those for the sequence $\{F(X_n), n \geq 1\}$. Since F(X) has an uniform distribution, it follows that the distribution of T_n, $n \geq 1$ does not depend on F. We will denote $T_{(n)}$ as the indices where the lower record values, $R_{(n)}$, occur. By our assumption $T_0 = T_{(0)} = 1$. The distribution of $T_{(n)}$ also does not depend on F. For more details, one may refer to Ahsanullah and Nevzorov (2001), Nevzorov (2000) and Arnold et al. (1998).

1.2. THE EXACT DISTRIBUTION OF RECORD VALUES

Many properties of the record value sequence can be expressed in terms of the function R(x), where $R(x) = -\ln \overline{F}(x)$, $0 < \overline{F}(x) < 1$ and $\overline{F}(x) = 1 - F(x)$. Here 'ln' is used for the natural logarithm. If we define $F_n(x)$ as the distribution function of R_n for $n \geq 0$, then we have

$F_0(x) = P[X_1 \leq x] = F(x)$ (1.2.1)

$F_1(x) = P[R_1 \leq x]$

$$= \int_{-\infty}^{x} \int_{-\infty}^{y} \sum_{i=1}^{\infty} (F(u))^{i-1} dF(u)\, dF(y)$$

$$= \int_{-\infty}^{x} \int_{-\infty}^{y} \frac{dF(u)}{1-F(u)} dF(y)$$

(1.2.2)

$$= \int_{-\infty}^{x} R(y) dF(y)$$

If F(x) has a density f(x), then the probability density function (pdf), $f_1(x)$, of R_1 is

$f_1(x) = R(x)\, f(x)$. (1.2.3)

The distribution function of R_2 is

$$F_2(x) = P(R_2 \leq x) = \int_{-\infty}^{x} \int_{-\infty}^{y} \sum_{i=0}^{\infty} (F(u))^{i} R(u) dF(u) dF(y)$$

$$= \int_{-\infty}^{x} \int_{-\infty}^{y} \frac{R(u)}{1-F(u)} dF(u) dF(y)$$

$$= \int_{-\infty}^{x} \frac{(R(u))^2}{\Gamma(3)} dF(u) \ . \quad (1.2.4)$$

The pdf $f_2(x)$ of R_2 is

$$f_2(x) = \frac{(R(x))^2}{\Gamma(3)} f(x), -\infty < x < \infty \ . \quad (1.2.5)$$

It can similarly be shown that the distribution function $F_{n-1}(x)$ of R_{n-1} is

$F_{n-1}(x) = P(R_{n-1} \leq x)$

$$= \int_{-\infty}^{x} f(u_n) du_n \int_{-\infty}^{u_n} \frac{f(u_{n-1})}{1-F(u_{n-1})} du_{n-1} \ldots \int_{-\infty}^{u_2} \frac{f(u_1)}{1-F(u_1)} du_1$$

$$= \int_{-\infty}^{x} \frac{[R(u)]^{n-1}}{\Gamma(n)} dF(u), \; -\infty < x < \infty. \tag{1.2.6}$$

This can be expressed as

$$F_{n-1}(x) = \int_{-\infty}^{R(x)} \frac{u^{n-1}}{\Gamma(n)} e^{-u} du, \; -\infty<x<\infty.$$

Therefore,

$$\overline{F}_{n-1}(x) = 1 - F_{n-1}(x) = \overline{F}(x) \sum_{j=0}^{n-1} \frac{(R(x))^j}{\Gamma(j+1)}$$

$$= e^{-R(x)} \sum_{j=0}^{n-1} \frac{(R(x))^j}{\Gamma(j+1)}$$

The pdf $f_{n-1}(x)$ of R_{n-1} is

$$f_{n-1}(x) = \frac{[R\ (x)]^{n-1}}{\Gamma(n)} f(x), \; -\infty<x<\infty. \tag{1.2.7}$$

A random variable X is said to be symmetric around zero if X and - X have the same distribution function. If f(x) is their density function, then f(x) = f(-x) for all x. Two random variables X and Y with cdfs $F_1(x)$ and $F_1^*(y)$ are said to be mutually symmetric i.e. $F_1(-x) = 1 - F_1^*(x)$ or equivalently if the corresponding pdfs $f_1(x)$ and $f_1^*(y)$ exist, then $f_1(-x) = f_1^*(x)$. If the sequence of i.i.d. random variables are symmetric around zero, then they are also mutually symmetric around zero but not conversely. For symmetric or mutually symmetric (around zero) sequence $\{X_n, n \geq 1\}$ of random variables, it can easily be seen that R_n is identically distributed as $- R_{(n)}$.

The joint pdf $f(x_0, x_1, x_2, \ldots, x_n)$ of the record values $R_0, R_1 R_2, \ldots, R_n$ is given by

$$f(x_0, x_1, \ldots, x_n) = r(x_0) r(x_1) \ldots r(x_{n-1}) f(x_n),$$
$$-\infty < x_0 < x_1 < \ldots < x_{n-1} < x_n < \infty, \tag{1.2.8}$$

where r(x) = , $0 < F(x) < 1$.

The function r(x) is known as hazard rate. The joint pdf of R_i and R_j, $1 \leq i < j$, is

$$f(x_i,x_j) = \frac{(R(x_i))^i}{\Gamma(i+1)} r(x_i) \frac{(R(x_j)-R(x_i))^{j-i-1}}{\Gamma(j-i)} f(x_j) \text{ , } -\infty < x_i < x_j < \infty . \quad (1.2.9)$$

In particular for i = 0 and j = n we have

$$f(x_0,x_n)=r(x_0)\frac{(R(x_n)-R(x_0))^{n-1}}{\Gamma(n)} f(x_n), \; -\infty<x_0<x_n<\infty .$$

Suppose we use the transformation $Y_1 = -\ln R_{i-1}$ and $Y_2 = \ln R_{i-1} / \ln R_{j-1}$, $1 \leq i < j$, then using (1.2.9), it can be shown that the pdf $f_2^*(y)$ of Y_2 is as follows:

$$f_2^*(y) = \frac{\Gamma(j)}{\Gamma(i)} \cdot \frac{1}{\Gamma(j-i)} \cdot y^{i-1} (1-y)^{j-i-1} \text{ , } 0 < y < \infty. \quad (1.2.10)$$

Thus Y_2 is distributed as Beta distribution with parameters i and j (i.e. B(i, j-i)). The mean and variance of Y_2 are

$$E(Y_2) = \text{ and } Var(Y_2) = \frac{i\ j}{(j+1)j^2} .$$

If we use the transformation $V_i = -\ln R_i$, then the joint pdf of V_i, i = 0,1,2,...,n is

$$f(v_0, v_1,..., v_n)= e^{-v_n} \text{ , } 0<v_0<v_1<\cdots<v_n<\infty . \quad (1.2.11)$$

The joint distribution of V_m and V_r is

$$f(v_m, v_r)=\frac{1}{\Gamma(m+1)} \cdot \frac{(v_r-v_m)^{r-m-1}}{\Gamma(r-m)} \cdot e^{-v_r} \text{ , } 0 < v_m < v_r < \infty$$

= 0 , otherwise.

Also we have

$$E(V_k^l) = \frac{1}{\Gamma(k+1)} \int_0^\infty t^l\ t^k e^{-t} dt = \frac{\Gamma(k+l+1)}{\Gamma(k+1)} .$$

Thus

$E(V_k) = k+1$ and $Var(V_k) = k+1$. The conditional pdf of $R_j | R_i = x_i$, $j > i$, is

$f(x_j \mid R_i = x_i) =$

$$= \frac{(R(x_j) - R(x_i))^{j-i-1}}{\Gamma(j-i)} \frac{f(x_j)}{\overline{F}(x_i)}, \quad -\infty < x_i < x_j < \infty. \tag{1.2.12}$$

For $j = i + 1$

$$f(x_{i+1} \mid R_i = x_i) = , \quad -\infty < x_i < x_{i+1} < \infty. \tag{1.2.13}$$

For $i > 0$, $0 \leq k < m$, the joint conditional pdf of R_{i+k} and $R_{i+m} \mid R_i$ is

$$f_{i+k,i+m}(x, y \mid R_i = z) = \frac{1}{\Gamma(m-k)} \frac{1}{\Gamma(k)} [R(y) - R(x)]^{m-k-1} [R(x) - R(z)]^{k-1} \frac{f(y) r(x)}{\overline{F}(z)},$$

$-\infty < z < x < y < \infty$.

The joint conditional distribution of $R_0, R_1, \ldots, R_{n-1}$ given R_n is given in the following Lemma.

Lemma 1.2.1.

Assume that F is continuous. The conditional joint distribution of $R_0, R_1, \ldots, R_{n-1}$ given $R_n = y$ is identical with the joint distribution of order statistics from a sample of size n from the cdf

$F_0(x \mid y) = R(x)/R(y)$, if $x < y$

$= 1$, if $x \geq y$.

Proof

The joint pdf $f_1(x_0, x_1, \ldots, x_{n-1} \mid y)$ of $R_0, R_1, \ldots, R_{n-1}$ given $R_n = y$, is

$$f_1(x_0, x_1, \ldots, x_{n-1} \mid y) = \frac{f(x_0, x_1, \ldots, x_{n-1}, y)}{f_n(y)}, \quad -\infty < x_0 < x_1 < \ldots < x_{n-1} < x_n < \infty$$

$$= \frac{r(x_0) r(x_1) \ldots r(x_{n-1}) f(y)}{f_n(y)}, \quad -\infty < x_0 < x_1 < \ldots < x_{n-1} < x_n < \infty$$

$$= \frac{n!\, r(x_0) r(x_1) \ldots r(x_{n-1})}{[R(y)]^n}, \ -\infty < x_0 < x_1 < \ldots < x_{n-1} < y < \infty.$$

Hence the result follows.

The marginal pdf of the nth lower record value can be derived by using the same procedure as that of the nth upper record value. If we use H(u) = -ln F(u), 0< F(u) < 1 and $h(u) = -\frac{d}{du} H(u)$, then

$$P(R(n) \le x) = \int_{-\infty}^{x} \frac{[H(u)]^n}{\Gamma(n+1)} dF(u) \qquad (1.2.14)$$

and the corresponding the pdf f(n) can be written as

$$f(n)(x) = \frac{[H(x)]^n}{\Gamma(n+1)} f(x). \qquad (1.2.15)$$

The joint pdf of $R_{(0)}$, $R_{(1),\ldots,}$ $R_{(n)}$ can be written as

$$f_{(0),(1),\ldots,(m)}(x_0, x_1, \ldots, x_m) = h(x_0) h(x_1) \ldots h(x_{m-1}) f(x_m), \ -\infty < x_m < x_{m-1} < \ldots < x_0 < \infty$$

= 0, otherwise. (1.2.16)

The joint pdf of R(r) and R(s) , is

$$f(r),(s)\ (x,y) = \frac{(H(x))^r}{\Gamma(r+1)} \frac{[H(y) - H(x)]^{s-r-1}}{\Gamma(s-r)} h(x) f(y), \ s > r \text{ and } -\infty < y < x < \infty. \qquad (1.2.17)$$

Using the transformations U= $H(R_{(s-1)})$ and W= $H(R_{s(-1))}/H(R_{(r-1))})$ in (1.2.17), it can be shown easily that W is distributed as B(r, s-r). Proceeding as in the case of upper record values, we can obtain the conditional pdfs of the lower record values.

Example 1.2.1.

Let us consider the exponential distribution with pdf f(x) as $f(x)=e^{-x}$, $0 \le x < \infty$ and the cumulative distribution function (cdf) F(x) as F(x) = 1- e^{-x}, $0 \le x < \infty$. Then R(x) = x and the pdf of R_n is given by

$$f_n(x) = \frac{x^n}{\Gamma(n+1)} e^{-x} ,$$

= 0, otherwise.

The joint pdf of R_{m-1} and R_{n-1}, $n > m \geq 1$, is

$$F_{m-1,n-1}(x,y) = \frac{x^{m-1}}{\Gamma(m)\,\Gamma(n-m)} (y-x)^{n-m-1} e^{-y} , 0 \leq x < y < \infty$$

= 0, otherwise.

The conditional pdf of $R_{n-1} \mid R_{m-1} = x$ is

$$f(y|R_{m-1} = x) = \frac{(y-x)^{n-m-1}}{\Gamma(n-m)} e^{-(y-x)} , 0 \leq x < y < \infty$$

= 0, otherwise.

Thus the conditional distribution of R_{n-1} given R_{m-1} is the same as the unconditional distribution of R_{n-m-1} for $n > m \geq 1$.

Example 1.2.2.

Suppose that the random variable X has the Gumbel distribution with pdf $f(x) = e^{-x} e^{-e^{-x}}, -\infty < x < \infty$. Let $F_{(n-1)}$ and $f_{(n-1)}$ be the cdf and pdf of $R_{(n-1)}$. It is easy to see that

$$F_{(n-1)}(x) = \int_{-\infty}^{\infty} \frac{e^{-nu}}{\Gamma(n)} e^{-e^{-u}} du$$

$$\text{and } f_{(n-1)}(x) = \frac{e^{-nx}}{\Gamma(n)} e^{-e^{-x}} , \qquad -\infty < x < \infty .$$

Let $f_{(m-1,n-1)}(x,y)$ be the joint pdf of $R_{(m-1)}$ and $R_{(n-1)}$, $0 < m < n$. Using (1.2.17), we get for the Gumbel distribution

$$f_{(m-1,n-1)}(x,y) = \frac{\left(e^{-y} - e^{-x}\right)^{n-m-1}}{\Gamma(n-m)} \frac{e^{-mx}}{\Gamma(m)} e^{-y} e^{-e^{-y}} , -\infty < y < x < \infty .$$

Thus the conditional pdf $f_{(n-1|m-1)}(y|x)$ of $R_{(n-1)}| R_{(m-1)}= x$ is given by

$$f_{(n-1|m-1)}(y|x)=\frac{(e^{-y}-e^{-x})^{n-m-1}}{\Gamma(n-m)}e^{-y}e^{-(e^{-y}-e^{-x})},\quad -\infty < y < x < \infty .$$

1.3. Distribution of Inter-Record Times

Let $\Delta_r = T_{r+1} - T_r$ and $\Delta_{(r)} = T_{(r+1)} - T_{(r)}$, r = 0,1, 2, ...,. We will call Δ_r and $\Delta_{(r)}$ as the upper and lower inter record times respectively. Since $T_0 = 1 = T_{(0)}$, we have $T_{r+1} = 1 + \Delta_1 + \Delta_2 + ... + \Delta_r$. Similarly $T_{(r+1)} = 1 + \Delta_{(1)} + \Delta_{(2)} +... + \Delta_{(r)}$.

Lemma 1.3.1.

For any n > 1, $P(\Delta_n < \infty) = 1 = P(\Delta_{(n)} < \infty)$.

Proof

$$P(\Delta_1=\infty)=P(\bigcap_{i=2}^{\infty} X_i \le X_1) \qquad (1.3.1)$$

$$= \lim_{m\to\infty} P(\max(X_2, X_2,..., X_m) \le X_1)$$

$$= \lim_{m\to\infty} \frac{1}{m} = 0.$$

It is obvious that $P(\Delta_{(1)} = \infty) = 0$.

Similarly it can be shown that for n $\geq$ 2, $P(\Delta_{(n)} = \infty) = 0$.

Theorem 1.3.1.

$$P(\Delta_n = k) = \sum_{i=0}^{k-1} \binom{k-1}{i} (-1)^i \frac{1}{(2+i)^n}.$$

Proof

$$P(\Delta_1 = k \mid X_1 = x_1) = P(R_1 = k+1 \mid X_1 = x_1) = .$$

Thus

$$P(\Delta_1 = k) = \int_{-\infty}^{\infty} (F(x_1))^{k-1}(1-F(x_1))f(x_1)\,dx_1 \tag{1.3.2}$$

$$= \frac{1}{k} - \frac{1}{k+1} = \frac{1}{k(k+1)}$$

Similarly it can be shown that

$$P(\Delta_n = k \mid R_{n-1} = x) = (F(x))^{k-1}(1-F(x)). \tag{1.3.3}$$

Thus

$$P(\Delta_n = k) = \int_{-\infty}^{\infty} (F(x))^{k-1}(1-F(x)) \frac{(-\ln(1-F(x))^{n-1}}{\Gamma(n)} f(x)\,dx$$

$$= \sum_{i=0}^{k-1} \binom{k-1}{i}(-1)^i \frac{1}{(2+i)^n} . \tag{1.3.4}$$

$$P(\Delta_n > k) = \int_0^{\infty} \sum_{j=k}^{\infty} (F(x))^{j-1}(1-F(x)) \frac{\{-\ln(1-F(x))\}^{n-1}}{\Gamma(n)} f(x)dx$$

$$= \sum_{i=0}^{k-1} \binom{k-1}{i}(-1)^i \frac{1}{(1+i)^n} \tag{1.3.5}$$

Proceeding similarly , the following theorem can be proved.

Theorem 1.3.2.

$$P(\Delta_{(n)} = k) = \sum_{i=0}^{k-1} \binom{k-1}{i}(-1)^i \frac{1}{(2+i)^n} .$$

For all $n \geq 1$, Δ_n and $\Delta_{(n)}$ are identically distributed. Their pdfs are independent of the parent distribution F(x). From (1.3.3) it follows that

$$E(\Delta_n \mid R_{n-1} = x) = \frac{1}{1-F(x)}$$

$$Var(\Delta_n \mid R_{n-1} = x) = \frac{F(x)}{(1-F(x))^2}$$

Similarly, we will have

$$E(\Delta_{(n)} \mid R_{(n-1)} = x) = \frac{1}{F(x)}$$

$$Var(\Delta_{(n)} \mid R_{(n-1)} = x) = \frac{1 - F(x)}{(F(x))^2}$$

Theorem 1.3.3.

$E(\Delta_n^{-1}) = n(S_{n+1} - 1)$, where

$$S_n = \sum_{j=1}^{\infty} (j)^{-n}.$$

Proof

$$E(\Delta_n^{-1}) = \frac{1}{\Gamma(n)} \int_{-\infty}^{\infty} \sum_{k=1}^{\infty} k^{-1} (F(x))^{k-1} (1-F(x))(-\ln(1-F(x)))^{n-1} f(x)\, dx$$

$$= \frac{1}{\Gamma(n)} \int_{-\infty}^{\infty} (F(x))^{-1} (1-F(x))(-\ln(1-F(x)))^{n} f(x)\, dx$$

$$= \frac{1}{\Gamma(n)} \int_{0}^{\infty} y^n e^{-y} (1-e^{-y})^{-1}\, dy$$

$$= n \sum_{k=1}^{\infty} (k+1)^{-(n+1)}.$$

Using the relation between S_{2n} and the Bernoulli numbers we can express $E(\Delta_n^{-1})$ for odd n in terms of the Bernoulli numbers.

Theorem 1.3.4.

$$P(\Delta_n \geq s) = \frac{1}{s}\sum_{j=1}^{s} P(\Delta_{n-1} \geq j).$$

Proof

$$P[\Delta_n \geq s] = \frac{1}{\Gamma(n)}\int_{-\infty}^{\infty} (1-F(x))\{-\ln(1-F(x))\}^{n-1}\sum_{k=s}^{\infty}(F(x))^{k-1} f(x)\, dx$$

$$= \frac{1}{\Gamma(n)}\int_{-\infty}^{\infty} (F(x))^{s-1}\{-\ln(1-F(x))\}^{n-1} f(x)\, dx$$

$$= \frac{1}{\Gamma(n)}\int_0^1 t^{s-1}\{-\ln(1-t)\}^{n-1}\, dt$$

$$= \frac{1}{s\,\Gamma(n-1)}\int_0^1 \frac{1-t^s}{1-t}\{-\ln(1-t)\}^{n-2}\, dt\,,\ n \geq 2,$$

$$= \frac{1}{s}\sum_{j=1}^{s} P(\Delta_{n-1} \geq j),\ n \geq 2. \qquad (1.3.7)$$

In particular

$$P(\Delta_1 > s) = \frac{1}{s+1},$$

and

$$P(\Delta_2 > s) = \frac{1}{s+1}\sum_{j=1}^{s} (1+j)^{-1}.$$

Theorem 1.3.5

$$E(\Delta_n^{\alpha}) = \infty \text{ , for } \alpha \geq 1 \text{ and } E(\Delta_n^{\alpha}) < \infty \text{ for } 0 < \alpha < 1.$$

Proof

Note that

$$\mathrm{E}(\Delta_{\mathrm{n}}) = \sum_{s=0}^{\infty} P[\Delta_n > s]$$

$$= \int_0^{\infty} \frac{y^{n-1}e^{-y}}{\Gamma(n)} \sum_{s=0}^{\infty} (1-e^{-y})^s \, dy \text{, by} \tag{1.3.5}$$

$$= \int_0^{\infty} \frac{y^{n-1}}{\Gamma(n)} dy = \infty,$$

and $E(\Delta_n^{\alpha}) > E(\Delta) = \infty$, $\alpha > 1$. For $0 < \alpha < 1$,

$$\mathrm{E}(\Delta_n^{\alpha}) = \frac{1}{\Gamma(n)} \int_{-\infty}^{\infty} \sum_{k=1}^{\infty} k^{\alpha} (\mathrm{F}(x))^{k-1} \overline{\mathrm{F}}(x) \{-\ln \overline{\mathrm{F}}(x)\}^{n-1} \mathrm{f}(x) \, dx$$

$$= \frac{1}{\Gamma(n)} \int_0^{\infty} e^{-2t} \, t^{n-1} \sum_{k=1}^{\infty} k^{\alpha} (1-e^{-t})^{k-1} \, dt.$$

Now

$$\sum_{k=1}^{\infty} k^{\alpha} (1-e^{-t})^{k-1} \leq \int_0^{\infty} (s+1)^{\alpha} e^{-s \ln(1-e^{-t})^{-1}} \, ds$$

$$= \int_1^{\infty} u^{\alpha} e^{-(u-1)\ln(1-e^{-t})^{-1}} \, du$$

$$\leq (1-e^{-t})^{-1} \int_0^{\infty} u^{\alpha} e^{-u \ln(1-e^{-t})^{-1}} \, du$$

$$= \left(1-e^{-t}\right)^{-1} \frac{\Gamma(1+a)}{\left\{-\ln\left(1-e^{-t}\right)\right\}^{a+1}}$$

$$\leq \left(1-e^{-t}\right)^{-1} \Gamma(1+a)\, e^{(1+a)t}.$$

Thus

$$E(\Delta_n{}^{\alpha}) \leq \frac{\Gamma(1+\alpha)}{\Gamma(n)} \int_0^{\infty} (1-e^{-t})^{-1} e^{-(1-\alpha)t} t^{n-1}\, dt .$$

Since $x \leq x(1-e^{-x})^{-1} \leq x+1$, we have

$$E(\Delta_n^{\alpha}) \leq \frac{\Gamma(1+\alpha)}{\Gamma(n)} \int_0^{\infty} e^{-(1-\alpha)x} (x+1)\, x^{n-2}\, dx$$

$$= \frac{\Gamma(1+\alpha)}{(1-\alpha)^n}\{1+\frac{1-\alpha}{n}\}.$$

1.4. The k- Records

Suppose that $X_1, X_2, \ldots$ is a sequence of independent and identically distributed random variables with cdf F(x) and pdf f(x). We define

$T_0(k) = k$

$T_{n-1}(k) = \min\{ j: j > T_{n-2}(k), X_j > X_{T_{n-2}(k)-k+1, T_{n-2}(k)} \}, n \geq 2,$

and

$T_{(0)}(k) = k$

$T_{(n-1)}(k) = \min\{ j: j > T_{(n-2)}(k), X_j < X_{k, T_{(n-2)}(k)} \}, n \geq 2,$

where $X_{j,m}$ is the jth order statistic in a random sample of size m. We call $T_m(k)$ and $T_{(m)}(k)$ as the mth upper and m^{th} lower - record indices respectively, m =0,1,2,.... We define $R_m(k)$ and $R_{(m)}(k)$ as the upper and lower k-record values, m = 0,1,2,.... If k =1, then these k- records will coincide with the records. The pdf of $R_m(k)$ is the same as the pdf of R_m from a random variable X with cdf as $1-(1-F(x))^k$. The lower record values of a sequence $\{X_n, n \geq 1\}$ are defined as the upper record values of the sequence $\{-X_n, n \geq 1\}$. Thus the lower k-record values of a sequence $\{X_n, n \geq 1\}$ are defined as the same as the upper k-record values of the sequence $\{-X_n, n \geq 1\}$.

Let $f_{n-1,k}(x)$ be the pdf of $R_{n-1}(k)$, then

$$f_{n-1,k}(x)=\frac{k^n}{\Gamma(n)}[-\ln(\overline{F}(x)]^{n-1}[\overline{F}(x)]^{k-1}f(x),\ n=1,2,\ldots,. \qquad (1.4.1)$$

The corresponding cdf is

$$F_{n-1,k}(x)=P(R_{n-1}(k)<x)=\int_{-\infty}^{x}\frac{k^n}{\Gamma(n)}[-\ln(\overline{F}(u)]^{n-1}[\overline{F}(u)]^{k-1}f(u)du$$

$$=\int_0^{-k\ln\overline{F}(x)}\frac{w^{n-1}e^{-w}}{\Gamma(n)}dw$$

$$=(\overline{F}(x))^k\sum_{j=0}^{n-1}\frac{(-k\overline{F}(x))^j}{\Gamma(j+1)} \qquad (1.4.2)$$

For m<n, the joint pdf $f_{m-1,n-1,k}(x,y)$ of $R_{m-1}(k)$ and $R_{n-1}(k)$ is defined as

$$f_{m-1,n-1,k}(x,y)=\frac{k^n}{\Gamma(m)\Gamma(n-m)}[-\ln\overline{F}(x)]^{m-1}[\ln\overline{F}(x)-\ln\overline{F}(y)]^{n-m-1}$$

$$\cdot\frac{f(x)}{\overline{F}(x)}[\overline{F}(y)]^{k-1}f(y),\ x<y \qquad (1.4.3)$$

The joint pdf $f_{0,1,2,\ldots,n}(x_0,x_1,x_2,\ldots,x_n)$ of $R_0(k), R_1(k),\ldots, R_n(k)$ is defined as

$$f_{0,1,2,\ldots,n}(x_0,x_1,\ldots,x_n)=k^n[\overline{F}(x_n)]^{k-1}f(x_n)\prod_{i=0}^{n-1}\frac{f(x_i)}{\overline{F}(x_i)},\ -\infty<x_0<x_1<\ldots<x_n<\infty,$$

$= 0$, otherwise. (1.4.4)

The pdf $f_{n|0,1,2,\ldots,n-1,k}$ of $R_n(k)$ given $R_0(k), R_1(k),\ldots,R_{n-1}(k)$ is

$$f_{n|0,1,2,\ldots n-1,k}(x_n|\ x_0,x_1,\ldots,x_{n-1})=\frac{k(\overline{F}(x_n))^{k-1}f(x_n)}{(\overline{F}(x_{n-1}))^k}$$

$$=f_{n|\,n-1,k}(x_n\,|x_{n-1}),$$

the pdf of $R_n(k)\,|R_{n-1}(k)=x_{n-1}$. Thus the sequence, $R_0(k), R_1(k),\ldots$ forms a Markov chain for any $k\geq 1$.

The joint pdf $f_{(0,1,2,\ldots,n)}(x_0, x_1, \ldots, x_{m-1})$ of $R_{(0)}, R_{(1)},\ldots,R_{(n)}$ is defined as

$f_{(0,1,2,\ldots,n)}(x_0, x_1, \ldots, x_n) =$

$$k^n [F(x_n)]^{k-1} f(x_n) \prod_{i=1}^{n-1} \frac{f(x_i)}{F(x_i)}, \; -\infty < x_n < x_{n-1} < \ldots < x_0 < \infty,$$

=0, otherwise. (1.4.5)

Example 1.4.1.

Let us consider the exponential distribution with pdf f(x) as $f(x)=e^{-x}$, $0\le x<\infty$ and the cumulative distribution function (cdf) F(x) as $F(x) = 1- e^{-x}$, $0 \le x < \infty$. We have ln (1-F(x)) =x and the pdf $f_{n-1,k}(x)$ of R_{n-1} is

$$f_{n-1,k}(x)=\frac{k^n}{\Gamma(n)} x^{n-1} e^{-kx}, x\ge 0, \; n=1,2,\ldots,$$

= 0, otherwise.

The joint pdf of R_{m-1} (k) and $R_{n-1}(k)$, n> m is

$$f_{m,n,k}(x,y) = \frac{k^n x^{m-1}}{\Gamma(m)\Gamma(n-m)} (y-x)^{n-m-1} e^{-ky}, \; 0 \le x < y < \infty$$

= 0, otherwise.

The conditional pdf of $R_{n-1}(k)$ given $R_{m-1}(k) = x$ is

$$f(y| R_{m-1}(k) = x) = \frac{k^{n-m} (y-x)^{n-m-1}}{\Gamma(n-m)} e^{-k(y-x)}, \; 0\le x < y < \infty,$$

= 0, otherwise.

Thus the conditional distribution of $R_{n-1}(k)-R_{m-1}(k)$ given $R_{m-1}(k)$ is the same as the unconditional distribution of $R_{n-m-1,k}$ for n > m.

We will define $\Delta_n(k) = T_n(k) - T_{n-1}(k)$, n =1,2,..., $\Delta_0(k) = k$, $\Delta_{(n)}(k) = T_{(n)}(k) - T_{(n-1)}(k)$, n =1,2,....,. If k =1, then $\Delta_n(k)$ and $\Delta_{(n)}(k)$ coincide with Δ_n and $\Delta_{(n)}$ respectively. We know that $E(T_n) = \infty$ for $n\ge 1$ but we will prove that $ET_n(k) < \infty$ for $n\ge 1$ and k>1.

$$P(\Delta_1(k)=r)=P(T_1(k)=k+r)=\int_{-\infty}^{\infty}[F(x)]^{r-1}\,k[\overline{F}(x)]^k\,f(x)\,dx$$

$$=\int_0^1 u^{r-1}\,k(1-u))^k\,du$$

$$=k\,\frac{\Gamma(k+1)\Gamma(r)}{\Gamma(k+r+1)}$$

$$E(T_1(k))=\sum_{r=1}^{\infty}k\frac{k!(r-1)!}{(k+r-1)!}=\frac{k^2}{k-1}.$$

$$P(\Delta_2(k)=r)=\int_{-\infty}^{\infty}[F(x)]^{r-1}\quad k[-k\,R(x)]\;[\overline{F}(x)]^k\,f(x)\,dx$$

$$=-k^2\int_0^1 u^k(1-u)^{r-1}\ln u\,du$$

$$=\sum_{j=0}^{r-1}\binom{r-1}{j}(-1)^j\frac{k^2}{(k+j+1)^2}$$

$$E(\Delta_2(k))=-k^2\int_0^1\sum_{r=1}^{\infty}r\,u^k(1-u)^{r-1}\ln u\,du=-k^2\int_0^1 u^{k-2}\ln u\,du=\frac{k^2}{(k-1)^2},\ k\geq 2.$$

Thus

$$E(T_2(k))=E(\Delta_1(k))+E(T_1(k))=\frac{k^2}{(k-1)^2}+\frac{k^2}{k-1}=\frac{k^3}{(k-1)^2}.$$

$$P(\Delta_n(k)=r)=\int_{-\infty}^{\infty}(F(x))^{r-1}\quad\frac{k[-k\,R(x)]^{n-1}}{\Gamma(n)}[\overline{F}(x)]^k\,f(x)\,dx$$

$$=\frac{k^n\int_0^1 u^k(1-u)^{r-1}(-\ln u)^{n-1}\,du}{\Gamma(n)}$$

$$=\sum_{j=0}^{r-1}\binom{r-1}{j}(-1)^j\frac{k^n}{(k+j+1)^n}\qquad(1.4.6)$$

$$E(\Delta_n(k)) = \sum_{r=1}^{\infty} \int_{-\infty}^{\infty} r(F(x))^{r-1} \frac{k[-k\,R(x)]^{n-1}}{\Gamma(n)} [\bar{F}(x)]^k f(x)\,dx$$

$$= \frac{k^n}{\Gamma(n)} \sum_{r=1}^{\infty} \int_0^1 r\,u^k (1-u)^{r-1} (-\ln u)^{n-1}\,du$$

$$= k^n \int_0^1 \frac{u^{k-2} (-\ln u)^n}{\Gamma(n)} du$$

$$= \frac{k^n}{(k-1)^n},\ k>1,\ n \geq 1. \quad (1.4.7)$$

Hence

$$E(T_n(k)) = E(\Delta_n(k)) + E(\Delta_{n-1}(k)) + \ldots + E(\Delta_1(k)) + k = \frac{k^{n+1}}{(k-1)^n},\ k>1,\ n \geq 0. \quad (1.4.8)$$

The probability of the event $\{\Delta_1(k)=k_1\} \cap \{\Delta_2(k)=k_2\}$ is

$P(\{\Delta_1(k)=k_1\} \cap \{\Delta_2(k)=k_2\})$

$$= \iint_{-\infty<x_0<x_1<\infty} (F(x_0))^{k_1-1} (F(x_1))^{k_2-1} \bar{F}(x_0)\bar{F}(x_1) r(x_0) f(x_1)$$

$dx_0\ dx_1$

$$= \int_{-\infty}^{\infty} k^2 (F(x_0))^{k_1-1} \bar{F}(x_0) r(x_0) u_0(x) dx_0,$$

where

$$u_0(x) = \sum_{j=0}^{k} \binom{k}{j} (-1)^j \frac{1}{k_2+1} (1-F(x_0))^{k_2+j}$$

$$= \int_{-\infty}^{\infty} k^2 \sum_{j=0}^{k} \binom{k}{j} (-1)^j u_1(x) f(x)\, dx ,$$

where

$$u_1(x) = \frac{1}{k_2 + j}\left((F(x))^{k_1 - 1} - (F(x))^{k_1 + k_1 + j - 1}\right)$$

$$= k^2 \left[\sum_{j=0}^{k} \binom{k}{j} (-1)^j \frac{1}{k_1 (k_1 + k_2 + j)} \right] \tag{1.4.9}$$

$$E\big(\Delta_n(k)(\Delta_n(k)+1)\big) = \sum_{r=1}^{\infty} \int_{-\infty}^{\infty} r(r+1)(F(x))^{r-1} \frac{k[-k\,R(x)]^{n-1}}{\Gamma(n)} [\overline{F}(x)]^k f(x)\, dx$$

$$= \int_{-\infty}^{\infty} \frac{2k[-k\,R(x)]^{n-1}}{\Gamma(n)} [\overline{F}(x)]^{k-3} f(x)\, dx$$

$$= \int_0^1 \frac{2k^n}{\Gamma(n)} t^{k-3} [-\ln t]^{n-1} dt = \left(\frac{k}{k-2}\right)^n , k > 2..$$

$$E(\Delta_n(k))^2 = \frac{2k^n}{(k-2)^n} - \frac{k^n}{(k-1)^n}, k > 2.$$

$$\operatorname{Var}\big(\Delta_n(k)\big) = \frac{2k^n}{(k-2)^n} - \frac{k^n}{(k-1)^n} - \frac{k^{2n}}{(k-1)^{2n}} k > 2.$$

The conditional probability distribution of $\Delta(n,k) \mid T_{n-1}(k)$ is geometric with

$$P(\Delta_n(k)=m \mid T_{n-1}(k) = x_m) = p\,(1-p)^{m-1},\ m=1,2,..., \tag{1.4.10}$$

where $p = 1 - F(x_m)$. Using the probability of inter-record times, we can easily obtain various probabilities of the record times.

$$P(T_1(k)=n_1) = P(\Delta_1(k) = n_1 - k)$$

$$= k\ \frac{k!(n_1-k-1)!}{n_1!}.$$

$$P\big(T_2(k)=n_2, T_1(k)=n_1\big)=P\big(\Delta_2=n_2-n_1, \Delta_1=n_1-1\big)$$

$$=k^2\left[\sum_{j=0}^{k}\binom{k}{j}(-1)^j\ \frac{1}{(n_1-k)(n_2-k+j)}\right]=\frac{k^2\ k!(n_2-k)!}{(n_1-k)(n_2-k)n_2!}$$

$$P\big(T_2(k)=n_2\mid T_1(k)=n_1\big)=P\big(\Delta_2(k)=n_2-n_1\mid\Delta_1(k)=n_1-1\big)$$

$$=k\frac{(n_1-k+1)(n_1-k+2)\ldots(n_2-k+1)}{(n_1+1)(n_1+2)\ldots n_2}\quad, n_2>n_1>k>1.$$

$$=k\frac{(n_2-k-1)!n_1!}{n_2!(n_1-k)!}\quad.$$

In general

$$\mathrm{P}\big(T_m(k)=n_m, T_{m-1}(k)=n_{m-1},\ldots,T_1(k)=n_1\big)$$

$$=\frac{k^{m-1}(n_m-1)!}{n_m!}\cdot\frac{k!}{(n_1-1)(n_2-k)(n_3-k)\ldots..(n_m-k)}$$

$$\mathrm{P}\big(T_m(k)=n_m\mid T_{m-1}(k)=n_1,\ldots,T_1(k)=n_1\big)=$$

$$\frac{k(n_{m-1}+1-k)(n_{m-1}+2-k)\ldots(n_m-k-1)}{(n_{m-1}+1)(n_{m-1}+2)\ldots(n_m-1)n_m}$$

$$=\frac{k\,n_{m-1}!(n_m-k-1)!}{(n_{m-1}-k)!n_m!}$$

$$=P\big(T_m(k)=n_m\mid T_{m-1}(k)=n_{m-1}\big).$$

Thus the sequence $\{T_n(k), k>1, n\geq 0\}$ forms a Markov chain.

1.5. Representations of Records

We present here representations of record values as sum on independent random variables. These relations will be useful for characterizations of distributions. Let $Y_1, Y_2, \ldots, Y_n, \ldots$ be a sequence of independent and identically distributed random variables with the cdf as $F_0(x) = 1-\exp(-x)$, $x>0$. Further suppose that $X_1, X_2, \ldots$, be a sequence i.i.d. r.v.'s with continuous cdf F. Then one has $R(X) \stackrel{d}{=} Y_1$. The following theorem (Bairamov and Ahsanullah (2000)) gives the representation of the nth record as sum on n independent random variables.

Theorem 1.5.1.

$R_{n-1} \stackrel{d}{=} g_F^{-1}\left(g_F(X_1) + g_F(X_2) + \ldots + g_F(X_n)\right)$, where $g_F(x) = R(x)$ and $g_F^{-1}(x) = F^{-1}(1-e^{-x})$. Using the result of example 1.2.1, we obtain

$P(R_{n-1} \le x) = P(Y_1+Y_2+\ldots+Y_n \le R(x))$

$= P(R(X_1) + R(X_2) + \ldots + R(X_n) \le R(x))$

$= P(g_F(X_1) + g_F(X_2) + \ldots + g_F(X_n) \le x)$

$= P(g_F^{-1}(g_F(X_1) + g_F(X_2) + \ldots + g_F(X_n)) \le x)$.

Therefore

$$R_{n-1} \stackrel{d}{=} g_F^{-1}\left(g_F(X_1) + g_F(X_2) + \ldots + g_F(X_n)\right). \qquad (1.5.1)$$

Using (1.5.1) we will give here the representation of nth record values of several well known distribution.

Example 1.5.1.

For the two parameter exponential distribution, $E(\mu, \sigma)$, with cdf F(x) as $F(x) = 1-e^{-(x-\mu)/\sigma}$, $x \ge 0$, we obtain

$$R_{n-1} \stackrel{d}{=} X_1 + X_2 + \ldots + X_n - (n-1)\mu$$

Example 1.5.2.

Consider the Weibull Distribution, W(α,β) with

$F(x) = 1 - e^{-\beta x^{\alpha}}$, $x \geq 0$, $\alpha > 0$, $\beta > 0$,

$= 0$, for $x < 0$.

Using Theorem 1.5.1, we obtain $R_{n-1} \overset{d}{=} (X_1^{\alpha} + X_2^{\alpha} + \ldots + X_2^{\alpha})^{\frac{1}{\alpha}}$.

Example 1.5.3.

Consider the Power Function Distribution, POW (δ,α,β) with cdf F(x) as

$$F(x) = \begin{cases} 0, x < 0 \\ 1 - \left(\dfrac{\beta - x}{\beta - \alpha}\right)^{\theta}, \alpha \leq x \leq \beta, \delta > 0, -\infty < \alpha < \beta < \infty, \\ 1, x \geq \beta. \end{cases}$$

For this distribution, the following representation is true

$$R_{n-1} \overset{d}{=} \beta - (\beta - \alpha)^{\frac{1}{\delta}} (\beta - X_1)(\beta - X_2)\ldots(\beta - X_n).$$

Example 1.5.4.

Consider the Pareto distribution, P(δ,α, β) with F(x) as

$$F(x) = \begin{cases} 1 - \left(\dfrac{\alpha + \beta}{x + \beta}\right)^{\delta}, x \geq \alpha \\ 0, x < \alpha. \end{cases}$$

It can easily be shown the following representation of R_n

$$R_{n-1} \overset{d}{=} \frac{1}{(\alpha + \beta)^{n-1}} (X_1 + \delta)(X_2 + \delta)\ldots(X_n + \delta) - \alpha$$

The following Theorem can be proved using the same procedure as given in Theorem 1.5.1 (for details see Ahsanullah and Nevzorov (2001, p. 242).

Theorem 1.5.2.

Suppose that $\{X_1, X_2, \ldots,\}$ and $\{Y_1, Y_2, \ldots, \}$ are two sequences of i.i.d. r.v.'s with continuous distribution functions F and H respectively. Let $R_0, R_1, \ldots,$ and $R^*_0, R^*_1, \ldots,$ be the corresponding record values. Then

$$\{R_0, R_1, \ldots,\} \overset{d}{=} \{G(H(R_0^*, R_1^*, \ldots,))\}, \tag{1.5.2}$$

where G is the inverse function of F.

Chapter 2

BOUNDS FOR THE MOMENTS OF FIRST RECORD STATISTICS

In this chapter we present the bounds for the expectations of the ordinary record statistics (k=1). Nagaraja (1978) provided the basic bound for $E(R_n)$ from general and symmetric distributions using Cauchy-Schwarz inequality. Other related bounds for ordinary records are given in Arnold et al. (1998) and Raqab (2000).

2.1. BOUNDS FOR FIRST RECORDS

We can write

$$E(R_n - \mu) = \int_0^1 (F^{-1}(u) - \mu)(f_n^*(u) - d)du,$$

$$\mu = E(X_1) \quad and \quad f_n^*(u) = \frac{[-\ln(1-u)]^n}{n!}, \tag{2.1.1}$$

for arbitrary constant d. Appling Cauchy-Schwarz inequality, we obtain

$$\frac{E(R_n - \mu)}{\sigma} \leq \left\{\binom{2n}{n} - 2d + d^2\right\}^{1/2} \tag{2.1.2}$$

Noting that the RHs of (2.1.2) is minimized when $d = 1$, we have

$$\frac{E(R_n - \mu)}{\sigma} \leq B_C(n),$$

where

$$B_C(n) = \left\{\binom{2n}{n} - 1\right\}^{1/2}, \tag{2.1.3}$$

(cf. Nagaraja (1978)). The bound (2.1.3) is attained for

$$F^{-1}(u) = \mu + \frac{\sigma}{\left\{\binom{2n}{n} - 1\right\}^{1/2}} \left\{ \frac{[-1n(1-u)]^n}{n!} - 1 \right\}, 0 < \mathrm{u} < 1,$$

and consequently

$$F(x) = 1 - e^{-n!\left\{B_C(n)(\frac{x-u}{\sigma})+1\right\}^{1/n}}, \; x \geq \mu - \frac{\sigma}{B_C(n)}.$$

For $n = 1$, $E(R_n) \leq \mu + \sigma$ and the inequality becomes equality iff

$$F(x) = 1 - e^{-\left[1+\frac{x-\mu}{\sigma}\right]}, x > \mu + \sigma.$$

If the parent distribution is symmetric about μ, then

$$\frac{E(R_n - u)}{\sigma} = \int_{/2}^{1} \left[F^{-1}(u) - \mu\right]\left[f_n^*(u) - f_n^*(1-u) - d\right] du.$$

Since the integrand is symmetric about 1/2,

$$E\frac{(R_n - \mu)}{\sigma} = \frac{1}{2}\int_0^1 \left[F^{-1}(u) - \mu\right]\left[f_n^*(u) - f_n^*(1-u) - d\right] du.$$

Applying Cauchy-Schwarz inequality, we obtain

$$E\frac{(R_n - \mu)}{\sigma} \leq \frac{1}{2}\left\{2\binom{2n}{n} + d^2 - \frac{2}{(n!)^2}\int_0^1 [\ln u \; \ln(1-u)]^n du\right\}^{1/2} \tag{2.1.4}$$

The RHS of (2.1.4) is minimized when $d = 0$. This leads to the following bound

$$E\frac{(R_n - \mu)}{\sigma} \leq B_C^S(n),$$

where

$$B_C^S(n) = \frac{1}{\sqrt{2}}\left\{\binom{2n}{n} - A_n\right\}^{1/2}, \tag{2.1.5}$$

and

$$A_n = \frac{1}{(n!)^2}\int_0^1 [\ln u \ \ln(1-u)]^n \, du \tag{2.1.6}$$

Nagaraja (1978) observed that the ratio of (2.1.3) and (2.1.5) tends to $1/\sqrt{2}$ as $n \to \infty$. A short cut formula is given for (2.1.6) in Arnold et al. (1998). Using the fact that

$$-\ln(1-u) = \sum_{i=1}^{\infty} \frac{u^i}{i},$$

One can rewrite A_n as

$$A_n = \frac{1}{n!}\sum_{j=n}^{\infty} \frac{c_j(n)}{(j+1)^{n+1}},$$

where $c_j(n)$ is the coefficient of u^j in $\left(\sum_{j=1}^{\infty} u^j/j\right)^n$. These coefficients can be generated in a recursive manner as

$$c_j(n) = \sum_{k=n-1}^{j-1} \frac{c_k(n-1)}{j-k},$$

(cf. Arnold et al., 1998, pp 70-71). One can also evaluate A_n and the bound easily using Mathematica via the codes

$f[u] = (Log[u] * Log[1-u]) \wedge n;$

$I =$ NIntegrate[f[u], $\{u,0,1\}$];

$C = (Gamma[n+1]) \wedge 2;$

$A = (1/c) * I$

Bound $= (0.5 \wedge 0.5) * (Binomial[2*n, n] - A) \wedge 0.5;$

Table 2.1 presents the values of bounds in (2.1.3) and (2.1.5) and the exact expected values from Normal, N(0,l). The numerical results are evaluated using Mathematica codes.

Table 2.1. Bounds of the Expected Values of Record Statistics in Terms of σ

n	Bound in (2.1.3)	Bound in (2.1.5)	Exact Value for N(0,1)
1	1.0000	0.9069	0.9032
2	2.2361	1.7269	1.4990
3	4.3589	3.1622	1.9687
4	8.3066	5.9161	2.3667
5	15.8430	11.2250	2.7174
6	30.3809	21.4942	3.0339
7	58.5747	41.4246	3.3244
8	113.4420	80.2185	3.5942
9	220.4972	155.9170	3.8471
10	429.8310	303.9370	4.0859

2.2. Extrapolation-Type Bound

Arnold et al. (1998) proposed this type of bound for record statistics. Here, we assume that the means of two upper records $(\alpha_j = E(R_j), j = r \ \ and \ \ s)$ are known. Without loss of generality, assume that $\mu = 0, \ \ and \ \ \sigma^2 = 1$. Let us write

$$\alpha_n = E(R_n) = \int_0^1 F^{-1}(u)\{f_n^*(u) - c[f_r^*(u) - \alpha_r F^{-1}(u)] - d[f_s^*(u) - \alpha_s F^{-1}(u)] - \lambda\} \ du,$$

where c, d and λ are arbitrary chosen constants. Applying Cauchy-Schwarz inequality to the above integral, we obtain

$$\alpha_n \le \binom{2n}{n} + c^2\left\{\binom{2r}{r} - \alpha_r^2\right\} + d^2\left\{\binom{2s}{s} - \alpha_s^2\right\}$$

$$+\lambda^2 - 2c\left\{\binom{r+n}{n} - \alpha_r \alpha_n\right\} - 2d\left\{\binom{s+n}{s} - \alpha_s \alpha_n\right\}$$

$$-2\lambda(1-c-d) + 2cd\left\{\binom{r+s}{s} - \alpha_r \alpha_s\right\}.$$

Setting $d = -c\alpha_r/\alpha_s$ and the optimal $\lambda = 1 - c(1-(\alpha_r/\alpha_s))$, we deduce the optimal $c = c^*$ as the

$$c^* = \frac{\left\{ q(r,n) - \frac{\alpha_r}{\alpha_s} q(s,n) \right\}^2}{q(r,r) - 2\frac{\alpha_r}{\alpha_s} q(r,s) + \frac{\alpha_r^2}{\alpha_s^2} q(s,s)},$$

where

$$q(i,j) = \binom{i+j}{i} - 1.$$

Therefore

$$\alpha_n = \{q(n,n) - c^*\}^{1/2} \tag{2.2.1}$$

As indicated in Arnold et al. (1998), the knowledge of α_r and α_s leads to an improvement over the bound in (2.1.3). For example, in the case of $N(0,1)$ with $r = 1$ and $s = 2$, $\alpha_1 = 0.9032$ and $\alpha_2 = 1.4990$. The bound in (2.2.1) for α_3 is 2.1229 and $\alpha_4 = 4.5194$, comparing with the respective bounds 4.3489 and 8.3066 obtained from (2.1.3). Analogous improvement over the bound in (2.1.5) can be achieved when the parent distribution is symmetric. The bound for symmetric distribution is given by

$$\alpha_n = \frac{1}{\sqrt{2}} \{q^*(n,n) - c^{**}\}^{1/2},$$

where

$$q^*(i,j) = \binom{i+j}{i} - \frac{1}{i!\,j!} \int_0^1 (-\ln u)^i (-\ln(1-u))^j \, du,$$

and

$$c^{**} = \frac{\left\{ q^*(r,n) - \frac{\alpha_r}{\alpha_s} q^*(s,n) \right\}^2}{q^*(r,r) - 2\frac{\alpha_r}{\alpha_s} q^*(r,s) + \frac{\alpha_r^2}{\alpha_s^2} q^*(s,s)}.$$

2.3. Hölder Inequality – Type Bound

All the preceding discussions have been concerned with establishing bounds for the mean of the record values based on Canchy-Schwarz inequality. Here, in this section, we present bounds based on Hölder's inequality. The method was used by Arnold (1985) in the context of order statistics and also sketched by Raqab (2000) for record statistics. We tacitly assume that the expectation of the parent distribution $\mu = E(X_1)$ and pth absolute central moment

$$\sigma_p^p = E|x-\mu|^p = \int_0^1 \left[F^{-1}(x)-\mu\right]^p dx, 1 \le p < \infty,$$

exist and are finite whenever they are used. Further, if the support of X is bounded, we denote the essential supremum of its deviation from the mean by

$$\sigma_\infty = ess\sup|x-\mu| = \sup_{x\in(0,1)} \left|F^{-1}(x)-\mu\right|.$$

The following result is a generation of that of Nagaraja (1978), who considered $p = 2$.

Theorem 2.1.

Let $p(1 < p < \infty)$ *be fixed and* $q = p/(p-1)$. *For any given* $n \ge 0$, *we have*

$$\frac{|ER_n - \mu|}{\sigma_p} \le B_H(n,p) \tag{2.3.1}$$

where

$$B_H(n,p) = \lambda^* \left[\frac{\lambda^*(n!)^{1/n}}{n}\right]^{\frac{p-1}{p}} \left\{\int_0^1 u^{-(n-1)/n}(1-u)^{p/(p-1)} \cdot e^{-[\lambda^*(n!)u]^{1/n}} du\right.$$

$$\left. + \int_1^\infty u^{-(n-1)/n}(n-1)^{p/(p-1)} e^{[\lambda^*(n!)u]^{1/n}} du\right\}^{(p-1)/p}, \tag{2.3.2}$$

where λ^* *is the solution to*

$$\int_0^1 u^{-(n-1)/n}(1-u)^{1/(p-1)} e^{-[\lambda^*(n!)u]^{1/n}} du$$

$$= \int_1^{\infty} u^{-(n-1)/n} (u-1)^{1/(p-1)} e^{[\lambda^*(n!)u]^{1/n}} du \qquad (2.3.3)$$

The bound $B_H(n, p)$ in (2.3.2) is attained for the location-scale family of distributions with

$$F(x) = 1 - e^{-n!\left\{\lambda^* + B_H(n,p)\left|\frac{x-\mu}{\sigma_p}\right| \operatorname{sgn}(x-\mu)\right\}},$$

$$\textit{for } -\lambda^* \big/ B_H(n,p) < \left|(x-\mu)/\sigma_p\right|^{p-1} \operatorname{sgn}(x-\mu) < \infty, \qquad (2.3.4)$$

where $\operatorname{sgn}(u) = 1$ for $u \geq 0$, -1 for $u < 0$.

Proof

The expected value of the nth record value can be written as

$$E(R_n - \mu) = \int_0^1 \left(F^{-1}(u) - \mu\right)\left(\frac{[-1n(1-u)]^n}{n!} - \lambda\right) du.$$

By applying the Hölder inequality (Mitrinovic (1970), p. 54), we obtain

$$E(R_n - \mu) \leq \sigma_p \left[\int_0^1 \left|\frac{[-1n(1-u)]^n}{n!} - \lambda\right|^{p/(p-1)} du\right]^{(p-1)/p}$$

$$= \sigma_p \left\{\int_0^{1-\bar{e}^{(\lambda n!)^{1/n}}} \left(\lambda - \frac{[-1n(1-u)]^n}{n!}\right)^{p/(p-1)} du + \int_{1-\bar{e}^{(\lambda n!)^{1/n}}}^1 \left(\frac{[-\ln(1-u)]^n}{n!} - \lambda\right)^{p/(p-1)} du\right\}^{(p-1)/p}$$

(2.3.5)

The optimal constant $\lambda = \lambda^*$ is obtained by differentiation of the RHS of the above inequality which is the one satisfying (2.3.3). The solution of (2.3.3) can be determined numerically using, for example, Newton-Raphson method. The bound so obtained is attained if the inverse cdf has the following form

$$F^{-1}(u) - \mu = \alpha\left[f_n(u) - \lambda^*\right]^{1/(p-1)} \operatorname{sgn}\left(f_n^*(u) - \lambda^*\right), \; 0 < u < 1, \qquad (2.3.6)$$

The condition $E|X_1-\mu|^p=\sigma_p^p$ forces $\alpha=\sigma_p / B_H^{p-1}(n,p)$.

Therefore

$$F^{-1}(u)-\mu=\frac{\sigma_p}{B_H^{p-1}(n,p)}\left|f_n^*(u)-\lambda^*\right|^{1/(p-1)}\operatorname{sgn}(f_n^*(u)-\lambda^*).$$

Straight forward algebra leads to the marginal cdf given in (2.3.4). This means that the bound is sharp and characterizes some special form of exponential model. In the case $p=2$ (2.3.4) reduces to the Weibull distribution (Nagaraja (1978). In the special case when $n=1$ (second record statistic), the bound in (2.3.1) reduces to

$$\frac{|ER_1-\mu|}{\sigma_p}\le B_H(n,1)$$

$$=\left(\lambda^*\right)^{(2p-1)/p}\left\{\int_0^1(1-u)^{p/(p-1)}\,\overline{e}^{\lambda^* u}\,du+\int_1^\infty(u-1)^{p/(p-1)}\,\overline{e}^{\lambda^* u}du\right\}^{(p-1)/p},$$

where λ^* satisfies

$$\Gamma\left(\frac{p}{p-1}\right)=\left(\lambda^*\right)^{1/(p-1)}\int_0^1 u^{p/(p-1)}\,e^{\lambda^* u}du.$$

A rough bound can be derived by setting $\lambda=0$ in (2.3.5). That is

$$\frac{|E(R_n-\mu)|}{\sigma_p}=B_H^*(n,p)\le\frac{1}{n!}\left\{\Gamma\left(1+\frac{np}{p-1}\right)\right\}^{(p-1)/p}. \qquad (2.3.7)$$

It is of interest to point out the sequence of record statistics $\{R_n, n=0,1\ldots,\}$ is just the sequence of $\{X_{T_n},\ n=0,1,2,\ldots,\}$. The maximum of a random sample of size n, $X_{n,n}$ is X_{T_k}, where $1\le k\le n$. This implies that the p-norm bounds obtained in (2.3.1) generalize the bounds on the moments of $X_{n,n}$ (see Arnold (1985)). Table 2.2 presents the upper bounds for $E(R_n-\mu)$ in (2.3.1) and (2.3.7) in various scale units. The bound in (2.3.2) results in a significant accuracy comparing with the one in (2.3.7). The discrepancy decreases as n gets large and the optimal choice of λ,λ^* converges to zero. Using the approximation $\Gamma(x)\approx\sqrt{2\pi}\,\overline{e}^{x}x^{x+1/2}$ for large x, it follows that both the bounds in (2.3.2) and (2.3.7) are approximated by

$$B_H(n,p)=\frac{1}{n^{3/(2p)}(2\pi)^{1/(2p)}}\left(\frac{p}{p-1}\right)^{n+[3(p-1)/2p]}.$$

Table 2.2. Bounds for the Expected Values of Records in Various Scale Units

	$p=2$		$p=3$		$p=5$	
n	$B_H(n,p)$	$B_H^*(n,p)$	$B_H(n,p)$	$B_H^*(n,p)$	$B_H(n,p)$	$B_H^*(n,p)$
1	1.0000	1.4142	0.8477	1.2090	0.7707	1.1051
2	2.2361	2.4495	1.5197	1.6510	1.2101	1.3069
3	4.3589	4.4721	2.2864	2.3321	1.5515	1.5763
4	8.3066	8.3666	3.3318	3.3472	1.9134	1.9193
5	15.8430	15.8745	4.8436	4.8486	2.3483	2.3496
6	30.3809	30.3974	7.0644	7.0660	2.8866	2.8868
7	58.5747	58.5833	10.3411	10.3416	3.5558	3.5559

Similarly we can derive p-norm bound for symmetric parent distribution. The respective bound is found to be

$$E\frac{(R_n-\mu)}{\sigma_p}\le B_H^S(n,p),$$

where

$$B_H^S(n,p)=\frac{(1/2)^p}{n!}\left\{\int_{1/2}^{1}\left[[-\ln(1-u)]^n-[-\ln u]^n\right]^{p/(p-1)}du\right\}^{(p-1)/p}. \quad (2.3.8)$$

Table 2.3. Bounds for the Expected Values of Records Based on Symmetric Parent in Various Scale Units

	$p=2$	$p=3$	$p=4$	$p=5$
n	$B_H^S(n,p)$	$B_H^S(n,p)$	$B_H^S(n,p)$	$B_H^S(n,p)$
1	0.9069	0.8027	0.7669	0.7488
2	1.7269	1.2925	1.1645	1.1037
3	3.1622	1.8498	1.5170	1.3685
4	5.9161	2.6566	1.9603	1.6705
5	11.2250	3.8483	2.5467	2.0455
6	21.4942	5.6083	3.3232	2.5131
7	41.4246	8.2081	4.3501	3.0956
8	80.2185	12.0511	5.7078	3.8202

For $p=2$, the bound in (2.3.8) reduces to the one derived in (2.1.5). From Table 2.3, it is evident that the p-norm bounds for symmetric parent distribution provide considerable improvement over the bound for a general distribution. As p increases, the bound decreases. The previous techniques applied for general distributions were based on decomposing the former integral in (2.3.5) into two integrals by finding the zero of the integrand function. This approach is not applied for k^{th} records and more numerical computations are needed for the optimal setting of λ.

Chapter 3

BOUNDS FOR THE MEAN OF K^{TH} RECORDS

In this section, we develop bounds for the moments of of kth record statistics from general continuous distributions based on combining the principle of Moriguti monotone approximation (Moriguti, 1953) with the Hölder inequalities. The variability behavior of the pdf of $R_{n,k}$, $f_{n,k}(x)$ is needed for the Moriguti approach to be applied. Raqab (1997) used the Moriguti approach combined with Cauchy-Schwarz inequality to develop optimal bounds for k^{th}records. The work was extended by combining the Moriguti monotone approximation with Hölder inequalities by Raqab and Rychlik (2002).

3.1. SCHWARZ INEQUALITY BASED BOUNDS

First let us recall Moriguti's (1953) approach in a simpler form. Suppose that a function h is integrable on $[a,b]$. Let $H(x) = \int_a^x h(t)dt,\ a \le x \le b$ and $\overline{H}$ be the convex minorant of H. Further let $\overline{h}$ be a non-decreasing version of derivative (e.g. right continuous) of $\overline{H}$. Obviously $\overline{h}$ is a non-decreasing function and constant in interval where $\overline{h} \ne h$. Function $\overline{h}$ represents the projection of h onto convex cone of non-decreasing functions in the Hilbert space $L^2([0,1], dx)$ (cf. Rychlik, 2001, pp. 14-16). Here we use Ph (projection function of h) as an alternative notation for $\overline{h}$.

Lemma 3.1.1.

For every non-decreasing function w on $[a,b]$ for which both the integrals (3.1.1) are finite, we have

$$\int_a^b w(x)h(x)\,dx \le \int_a^b w(x)\,Ph(x)\,dx. \tag{3.1.1}$$

The equality in (3.1.1) holds iff w is constant in every interval constant in the set where $\overline{H} \neq H$. The expected value of $R_{n,k}$ can be written as

$$E(R_{n,k}) = \int_0^1 F^{-1}(x)\, f^*_{n,k}(x)\, dx\,,$$

where

$$f^*_n(x) = k^{n+1}\frac{[-\ln(1-x)]^n}{n!}(1-x)^{k-1},\ 0 < x < 1,$$

is the density of the nth value of kth record statistics from a standard uniform distribution U(0, 1). Its respective cdf can be rewritten using as

$$F^*_{n,k}(x) = 1-(1-x)^k \sum_{j=0}^{n} \frac{[-\ln(1-x)]^j}{j!}. \tag{3.1.2}$$

For simplicity we write the expected value of $R_{n,k}$ as follows:

$$E(R_{n,k}) = \int_0^\infty F^{-1}\left(1-\bar{e}^{y}\right) g_{n,k}(y)\, \bar{e}^{y} dy, \tag{3.1.3}$$

where

$$g_{n,k} = \frac{k^{n+1}}{n!} x^n\ e^{-(k-1)x},\ x > 0,$$

is a density function with respect to the exponential measure on the positive half axis. Its respective cdf is

$$G_{n,k}(x) = 1-e^{-nx}\sum_{j=0}^{\infty}\frac{(kx)^j}{j!}, \tag{3.1.4}$$

which is the distribution function of $\Gamma(n+1, x)$ random variable. Therefore

$$E(R_{n,x} - \mu) = \int_0^\infty \left[F^{-1}(1-\bar{e}^{y}) - \mu\right]\left[g_{n,k}(y) - d\right] \bar{e}^{y} dy, \tag{3.1.5}$$

for arbitrary constant d. Grudzien and Szynal (1985) have shown that

$$E(R_{n,k}-\mu)\le B_C(n,k)=\left\{\frac{k^{2n+2}}{(2k-1)^{2n+1}}\binom{2n}{n}-1\right\}^{1/2}. \quad (3.1.6)$$

Clearly for $k>1$, the bound $B_C(n,k)$ is decreasing in n if $n<2k^2/(4k-1)$ and it is increasing if $n>2k^2/(4k-1)$. For $k=1$, $g_{n,k}(y)$ is a monotonic function and the bound is attained iff $F(x)$ takes the form given in (2.3.4). In general, function $g_{n,k}$ is not monotonic except for the case $n=0$ (parent r.v.) and then the bound obtained by Grudzien and Szynal (1985) is not attainable. Further $g_{n,k}(y)$ is increasing, decreasing with the maximum at $n/(k-1)$. It follows that the anti-derivative $G_{n,k}(y)$ in (3.1.4) is convex increasing on [0, n/(k-1)) and concave increasing on $(n/(k-1),\infty)$. The greatest convex minorant $\overline{F}^*_{n,k}$ is linear in $[1\text{- e}^{-\beta},1]$ for β satisfying

$$1-G_{n,k}(\beta)=g_{n,k}(\beta)\,\bar{e}^{\beta}. \quad (3.1.7)$$

The root β of (3.1.7) can be found numerically using for example Newton-Raphson Method. The projection of $f^*_{n,k}$ onto the family of non-decreasing functions in $L^2([0,1],dx)$ has the form

$$Pf^*_{n,k}(x)=f^*_{n,k}(x)\left(\min\{x,1-\bar{e}^{\beta}\}\right)$$

Also

$$Pg_{n,k}(x)=g_{n,k}\left(\min\{x,\beta\}\right),$$

is the analogous projection of $g_{n,k}$ in $L^2\left(\left[0,\infty\right],\bar{e}^{y}dy\right)$. It follows that

$$E(R_{n,k}-\mu)\le\int_0^\infty\left(F^{-1}(1-\bar{e}^{y})-\mu\right)\left[Pg_{n,k}-d\right]e^{-y}dy$$

$$\le\left\{\int_0^\infty\left(Pg_{nk}(y)-d\right)^2\bar{e}^{y}dy\right\}^{1/2}\sigma. \quad (3.1.8)$$

The RHS is minimized by choosing $d=1$ and the optimal bound is given by

$$B_{PC}(n,k)=\left\{\int_0^\infty(Pg_{n,k}(y)-1)^2\,\bar{e}^{y}dy\right\}^{1/2}\sigma$$

$$= \left\{ \int_0^{\beta} \left(g_{n,k}(y) - 1\right)^2 \bar{e}^{y}\, dy + \left(g_{n,k}(\beta) - 1\right)^2 \bar{e}^{\beta} \right\}^{1/2} \sigma. \tag{3.1.9}$$

The later inequality in (3.1.8) becomes equality iff $\overline{F}^{1}\left(1-\bar{e}^{y}\right)-\mu=\alpha\left(Pg_{n,k}-1\right)$ for some $\alpha > 0$, which satisfies the former equality as well. The condition $\sigma^2 = \int_0^{\infty} \left[F^{-1}(1-\bar{e}^{y}) - \mu\right]^2 \bar{e}^{y} dy$ forces $\alpha = \sigma / B_{PC}(n,k)$. Consequently, the distribution function of the location-scale family for which the bound $B_{PC}(n,k)$ in (3.1.9) is attained have the form

$$F(x) = \begin{cases} 0 & if\ x \le \mu - \dfrac{\sigma}{B_{PC}(n,k)} \\ f_{n,k}^{*-1}\left(1 + B_{PC}(n,k)\left(\dfrac{x-\mu}{\sigma}\right)\right) & if\ \mu - \dfrac{\sigma}{B_{PC}(n,k)} < x < \mu + \left[\dfrac{\left(g_{n,k}(\beta)-1\right)\sigma}{B_{PC}(n,k)}\right], \\ 1 & if\ x \ge \mu + \left[\dfrac{\left(g_{n,k}(\beta)-1\right)\sigma}{B_{PC}(n,k)}\right]. \end{cases} \tag{3.1.10}$$

The distribution function in (3.1.10) involves the inverse of a smooth function and a jump of height $\bar{e}^{\beta}$ at the right end.

Remark 3.1.1.

The equation (3.1.7) has explicit solutions for $n=1$ and $n=2$. If $n=1$, the root is obtained to be

$$\beta = 1/\left(k(k-1)\right), k = 2,3,...,$$

If $n=2$, then the root is found to be

$$\beta = \left\{1 + \left[1 + 2(k-1)\right]^{1/2}\right\} / k(k-1),\ k = 2,3,...,$$

3.2. Hölder Based Bounds

In this section, we present bounds for the kth record statistics of general population. The bounds are expressed in the scale units σ_p determined by the pth central absolute moments described in Section 2.3. Using the Hölder inequality, we obtain from (3.1.4)

$$E(R_{n,k}-\mu)\le \left\|F^{-1}-\mu\right\|_p \left\|g_{n,k}-d\right\|_q = \left\|g_{n,k}-d\right\|_q \sigma_p, \qquad (3.2.1)$$

where $\|g\|_p = \left[\int_0^\infty g^p(x)\ \bar{e}^x dx\right]^{1/p}$, denotes the norm of $g\in L^p([0,\infty),\ \bar{e}^y\,dy)$ and $\|g\|_q$ is defined analogously for the conjugate exponent $q=p/(p-1)$. Putting $d=0$, we obtain a simple analytic bound

$$B_H(n,k,p) = \left\|g_{n,k}\right\|_q = \frac{k^{n+1}}{n!}\left\{\frac{\Gamma\left(1+\frac{np}{p-1}\right)}{\left[1+\frac{(k-1)p}{p-1}\right]^{(np/(p-1))+1}}\right\}^{(p-1)/p}. \qquad (3.2.2)$$

Bound (3.2.2) is not optimal since the constant $d=0$ does not minimize the norm and the equality of the Hölder inequality in (3.2.1) can not be attained for general k. For $k=1$, the equality in (3.2.1), can be achieved iff

$$F^{-1}(1-\bar{e}^y)-\mu=\phi_p(g_{n,k}(y)-d),$$

for monotone function $\phi_p(y)=|y|^{1/(p-1)}\operatorname{sgn}(y)$ and properly chosen $d=d^*$, being the solution to (2.3.3). The extreme distribution is defined in (2.3.4). If $k>1$, then both $g_{n,k}(y)$ and $\phi_p(g_{n,k}(y)-c)$ are not monotonic functions except for the case $n=0$ (parent r.v.). Therefore the bound in (3.2.1) is not attainable and we use the Moriguti (1953) approximation based on the greatest convex minorant for improving it. Arguments similar to those in Section 1, we conclude

$$E(R_{n,k}-\mu)\le \int_0^\infty \left[F^{-1}(1-\bar{e}^y)-\mu\right]\left[Pg_{n,k}(y)-d\right]\bar{e}^y dy$$

$$\le \left\|Pg_{n,k}(y)-d\right\|_q \sigma_p, \qquad (3.2.3)$$

for arbitrary read *d*. We now minimize the norm in (3.2.3) with respect to $d=g_{n,k}(c),\ c\in(0,\beta)$ where β is defined in (3.1.7). We have

$$\left\|Pg_{n,k}(y)-g_{n,k}(c)\right\|_q^q = \int_0^c\left[g_{n,k}(c)-g_{n,k}(y)\right]^q \bar{e}^y dy$$

$$+\int_0^{\beta}\left[g_{n,k}(y)-g_{n,k}(c)\right]^q\ \bar{e}^{y}\ dy+\left[g_{n,k}(\beta)-g_{n,k}(c)\right]^q\ \bar{e}^{\beta}. \tag{3.2.4}$$

Differentiating of the RHS of (3.2.4) and equating the result to 0 leads to

$$\int_0^{c}\left[g_{n,k}(c)-g_{n,k}(y)\right]^{q-1}\bar{e}^{y}\ dy=\int_c^{\beta}\left[g_{n,k}(y)-g_{n,k}(c)\right]^{q-1}\ \bar{e}^{\beta}\ dy$$

$$+\left[g_{n,k}(\beta)-g_{n,k}(c)\right]^{q-1}\ \bar{e}^{\beta}. \tag{3.2.5}$$

One can easily show that there is a unique solution c^* of equation (3.2.5) that minimizes (3.2.4). Therefore, the optimal bound on $E(R_{n,k}-\mu)/\sigma_p$ is given by

$$B_{PH}(n,k,p)=\left\|Pg_{n,k}(y)-g_{n,k}(c^*)\right\|_q, \tag{3.2.6}$$

that can be calculated from (3.2.4) with $c=c^*$. The former inequality in (3.2.3) becomes equality if $F^{-1}(1-\bar{e}^{y})$ is constant on (β,∞). The latter one is attained iff

$$F^{-1}(1-\bar{e}^{y})-\mu=\lambda\left[Pg_{n,k}(y)-g_{n,k}(c^*)\right]^{q/p}\operatorname{sgn}\left(Pg_{n,k}(y)-g_{n,k}(c^*)\right),$$

for some $\lambda\ge 0$. The condition $\int_0^1\left|\bar{F}^1(x)-\mu\right|^p dx=\sigma_p^p$ forces

$$\lambda=\frac{\sigma_p}{\left\|Pg_{n,k}-g_{n,k}(c^*)\right\|_q^{q/p}}=\frac{\sigma_p}{\left[B_{PH}(n,k,p)\right]^{q/p}}.$$

It follows that the distribution functions of the location-scale family for which the bounds are attained have the form

$$F(x)=\begin{cases}0, & \text{if } x\le\beta,\\ f_{n,k}^{*-1}\left(g_{n,k}(c^*)+\operatorname{sgn}(x-\mu)B_{PH}(n,k,p)\left|\dfrac{x-\mu}{\sigma_o}\right|^{p/q}\right) & \text{if } \beta<x<\gamma,\\ 1, & \text{if } x\ge\gamma,\end{cases} \tag{3.2.7}$$

where

$$\beta = \mu - \sigma_p \left[\frac{g_{n,k}(c^*)}{B_{PH}(n,k,p)} \right]^{q/p}, \text{ and } \gamma = \mu + \sigma_p \left[\frac{g_{n,k}(\beta) - g_{n,k}(c^*)}{B_{PH}(n,k,p)} \right]^{q/p}.$$

Remark 3.2.1.

In the special case of ordinary RS's, function $g_{n,1}(\cdot)$ is strictly increasing on $(0,\infty)$ and the greatest convex minorant principle does not need to be used, and we can put $\beta = \infty$. In this case, the bound reduces to (2.3.2). For $p = 2$, c^* is simplified to $g_{n,k}^{-1}(1)$, and the bound (3.2.6) reduces to the bound given in (3.1.8).

Now we consider the extreme case $p = 1$ with $k > 1$. In this case we have

$$E(R_{n,k} - \mu) \le \sup_{x \in (0,1)} \left| Pf_{n,k}^*(x) - d \right| \int_0^1 \left| F^{-1}(x) - \mu \right| dx = \left\| Pf_{n,k}^* - d \right\|_\infty \sigma_1,$$

where, by monotonicity of $Pf_{n,k}^*$, the constant $d = d^*$ minimizing the norm is

$$d^* = \frac{\lfloor Pf_{n,k}^*(0) + Pf_{n,k}^*(1) \rfloor}{2} = \frac{g_{n,k}(\beta)}{2}.$$

The optimal bound on $E(R_{n,k} - \mu)/\sigma_p$ has the form

$$B_{PH}(n,k,1) = \left\| Pf_{n,k}^* - c^* \right\|_\infty = \frac{Pf_{n,k}^*(1) - Pf_{n,k}^*(0)}{2}$$

$$= \frac{g_{n,k}(\beta)}{2} = \frac{k^{n+1}\beta^n}{2n!\, e^{(k-1)\beta}}. \tag{3.2.8}$$

It is easy to verify that this is attained in the limit by the sequence of three-point distribution supported at $\mu - j\sigma_1/2$, μ, and $\mu + \sigma_1 e^{\beta}/2$ with respective probabilities $1/j$, $1 - e^{-\beta} - 1/j$, and $e^{-\beta}$, as $j \to \infty$. Note that now the rough bound (3.2.2) has the form

$$B_H(n,k,1) = \sup_{y \in (0,\infty)} \left| g_{n,k}(y) \right| = g_{n,k}\left(\frac{n}{k-1} \right) = \frac{k^{n+1} n^n}{n!(k-1)^n e^n},$$

and we have $B_{PH}(n,k,1) < B_H(n,k,1)$ for all $n \geq 1$ and $k > 1$. If $k = 1$, then both $B_H(n,k,1)$ and $B_{PH}(n,k,1)$ are infinite.

If $p = \infty$ and $q = 1$, we have (3.2.6) with the optimal constant

$$d^* = f^*_{n,k}(1/2) = g_{n,k}(\ln 2) = \frac{k^{n+1}(\ln 2)^n}{n!\ 2^{k-1}}.$$

The best bound is

$$B_{PH}(n,k,\infty) = \left\| Pg_{n,k} - d^* \right\|_1$$

$$= \begin{cases} g_{n,k}(\beta) - 1, & \text{if } \beta \leq \ln 2, \\ 1 - 2G_{n,k}(\ln 2), & \text{if } \beta \geq \ln 2, \end{cases}$$

$$= \begin{cases} \frac{k^{n+1}}{n!}\beta^n e^{-(k-1)\beta} - 1, & \text{If } \beta \leq \ln 2, \\ 1 - \frac{1}{2^{k-1}}\sum_{j=0}^{n} \frac{(k \ln 2)^j}{j!}, & \text{If } \beta \geq \ln 2, \end{cases} \tag{3.2.9}$$

while

$$B_H(n,k,\infty) = G_{n,k}(\infty) = 1.$$

If $\beta < \ln 2$, then bound (3.2.9) is attained if $F^{-1}(x) - \mu = -\sigma_\infty$ for $x < 1 - e^{-\beta}$, and is a constant otherwise. The first moment condition implies that F defines the two-point distribution supported on $\mu - \sigma_\infty$ and $\mu + \left(e^{\beta} - 1\right)\sigma_\infty$ with probabilities $1 - e^{\beta}$ and e^{β}, respectively. The latter bound in (3.2.9) is attained if

$$F^{-1}(x) - \mu = \operatorname{sgn}(x - 1/2)\sigma_\infty,$$

which defines the distribution function symmetrically distributed at $\mu \pm \sigma_\infty$.

Table 3.1 includes results for k^{th} records with $k = 2, 3, 4, 5$. The solution of (3.1.7) is determined numerically by means of the Newton-Raphson method. The simplified form of (3.1.7) is given by

$$\sum_{j=0}^{n} \frac{k^j \beta^j}{j!} = \frac{k^{n+1}\beta^n}{n!}.$$

Table 3.1. Bounds on the Expectations of kth Records in Various Scale Units

		$p=1$		$p=2$		$p=3$		$p=5$		$p=\infty$
k	n	B_H (n,k,1)	B_{PH} (n,k,1)	B_H (n,k,2)	B_{PH} (n,k,2)	B_H (n,k,3)	B_{PH} (n,k,3)	B_H (n,k,5)	B_{PH} (n,k,5)	B_{PH} (n,k,∞)
2	1	1.4715	0.6065	1.0887	0.3451	1.0502	0.3003	1.0269	0.2662	0.2131
	2	4.3307	0.9521	1.2571	0.7361	1.1472	0.7095	1.0795	0.6931	0.6736
	3	21.508	1.6243	1.5301	1.1451	1.2965	1.0494	1.1574	0.9826	0.8956
	4	150.04	2.8843	1.9083	1.6171	1.4886	1.3516	1.2526	1.1788	0.9726
	5	1347.6	5.2443	2.4139	2.1910	1.7251	1.6501	1.3631	1.3334	0.9939
	6	14803	9.6862	3.0815	2.9100	2.0112	1.9705	1.4887	1.4771	0.9988
	7	$1.92x10^{5}$	18.092	3.9592	3.8269	2.3548	2.3328	1.6299	1.6255	0.9998
	8	$2.88x10^{6}$	34.078	5.1113	5.0092	2.7658	2.7540	1.7879	1.7862	1.0000
	9	$4.90x10^{7}$	64.609	6.6230	6.5442	3.2566	3.2502	1.9642	1.9635	1.0000
	10	$9.30x10^{8}$	123.14	8.6071	8.5462	3.8419	3.8383	2.1602	2.1599	1.0000
3	1	1.6654	0.5374	1.1384	0.2137	1.0795	0.1529	1.0431	0.1147	0.0748
	2	3.6540	0.6677	1.1831	0.4807	1.1056	0.4389	1.0573	0.4029	0.3354
	3	13.611	0.8803	1.2960	0.7336	1.1714	0.7136	1.0931	0.7008	0.6850
	4	71.211	1.1960	1.4548	0.9920	1.2609	0.9529	1.1408	0.9234	0.8798
	5	479.69	1.6551	1.6561	1.2692	1.3698	1.1642	1.1971	1.0815	0.9608
	6	3951.9	2.3188	1.9027	1.5757	1.4972	1.3601	1.2607	1.2004	0.9889
	7	38493	3.2781	2.2002	1.9217	1.6435	1.5516	1.3310	1.2995	0.9972
	8	$4.82x10^{5}$	4.6660	2.5564	2.3178	1.8097	1.7475	1.4079	1.3909	0.9994
	9	$5.51x10^{6}$	6.6771	2.9813	2.7756	1.9976	1.9549	1.4915	1.4818	0.9999
	10	$7.85x10^{7}$	9.5966	3.4870	3.3089	2.2094	2.1793	1.5818	1.5759	1.0000
4	1	1.9620	0.5192	1.2218	0.1554	1.1288	0.0978	1.0702	0.0670	0.0384
	2	3.8495	0.5936	1.2092	0.3674	1.1210	0.3045	1.0658	0.2535	0.1872
	3	12.746	0.7103	1.2916	0.5620	1.1517	0.5258	1.0826	0.4926	0.4206
	4	59.276	0.8706	1.3487	0.7512	1.2021	0.7333	1.1100	0.7211	0.7038
	5	354.92	1.0829	1.4623	0.9432	1.2664	0.9190	1.1443	0.9013	0.8745
	6	2599.2	1.3608	1.6000	1.1437	1.3423	1.0855	1.1839	1.0359	0.9534
	7	25504	1.7227	1.7621	1.3573	1.4287	1.2384	1.2281	1.1383	0.9846
	8	$2.24x10^{5}$	2.1933	1.9498	1.5883	1.5256	1.3834	1.2762	1.2210	0.9954
	9	$2.55x10^{6}$	2.8051	2.1656	1.8406	1.6330	1.5254	1.3282	1.2932	0.9988
	10	$3.22x10^{7}$	3.6007	2.4123	2.1185	1.7513	1.6687	1.3839	1.3605	0.9997
5	1	2.2993	0.5117	1.3095	0.1223	1.1799	0.0702	1.0981	0.0450	0.0234
	2	4.2292	0.5617	1.2600	0.3015	1.1509	0.2304	1.0822	0.1799	0.1233
	3	13.127	0.6396	1.2781	0.4644	1.1613	0.4093	1.0878	0.3584	0.2792
	4	57.236	0.7429	1.3283	0.6187	1.1905	0.5876	1.1037	0.5578	0.4858
	5	321.29	0.8737	1.4002	0.7708	1.2318	0.7544	1.1260	0.7428	0.7242
	6	2205.8	1.0364	1.4895	0.9245	1.2823	0.9064	1.1529	0.8936	0.8746
	7	17905	1.2372	1.5948	1.0831	1.3405	1.0445	1.1834	1.0106	0.9491
	8	$1.67x10^{5}$	1.4841	1.7158	1.2489	1.4058	1.1718	1.2170	1.1018	0.9814
	9	$1.78X10^{6}$	1.7872	1.8527	1.4242	1.4779	1.2914	1.2534	1.1753	0.9938
	10	$2.11x10^{7}$	2.1590	2.0064	1.6111	1.5567	1.4065	1.2924	1.2379	0.9981

Using (3.2.8) and (3.2.9), we compute the values of bounds in the extreme cases $p=1$ and $p=\infty$, respectively. Otherwise, we determine the constants c^* that minimize the norms $\left\|Pg_{n,k} - g_{n,k}(c^*)\right\|_q$.

In Table 3.1, each optimal bound $B_{PH}(n,k,p)$ is compared with the respective rough one $B_H(n,k,p)$ for $1 \le n \le 10$, $k = 2, 3, 4, 5$ and $p = 1, 2, 3, 5, \infty$. Note that

$B_H(n,k,\infty)=1$ for all n and k. The numerical results show that the direct application of the Hölder inequality without the Moriguti modification generally results in a significant loss of accuracy in evaluating small values of kth records. The difference becomes smaller as n increases. The exceptional case is $p=1$ for which $B_H(n,k,p)$ become totally awkward for large n. We clearly observe the increase and decrease of our bounds with respect to n and k, correspondingly, which reflects the relations $R_{n,k+1} \le R_{n,k} \le R_{n+1,k}$ for the records themselves. We also note that the bounds decrease as p increases. Actually, if we treat σ_p as a general scale parameter and increase p, we restrict ourselves to narrower classes of distributions for which p^{th} moments exist, and the bounds in the narrower classes become tighter.

3.3. Steffensen-Type Inequalities

The Steffensen inequalities (Mitrinovic, 1972) are applied to establish moments bounds for RS's based on independent identically distributed variables. Let us present the following Lemma:

Lemma 3.3.1.

Let w_1 and w_2 be integrable functions defined on $[a,b]$ so that w_1 is non-decreasing and $0 \le w_2(t) \le 1$ for $t \in [a,b]$. If $\lambda = \int_0^1 w_2(t)dt$, then the following inequalities hold

$$\int_a^{a+\lambda} w_1(t)\ dt \le \int_a^b w_1(t)\,w_2(t)\ dt \le \int_{b-\lambda}^b w_1(t)\ dt. \tag{3.3.1}$$

The first inequality becomes equality iff $w_2(.) = I_{[a,a+\lambda]}(.)$ a.e. on $[a,b]$ or w_1 = constant a.e. on the set $\{x \in [a,a+\lambda] : w_2(x) < 1\}$ or $\{x \in [a+\lambda,\ b] : w_2(x) > 0\}$. While the second inequality becomes equality iff $w_2(.) = I_{[b-\lambda,b]}(.)$ a.e. in $[a,b]$ or w_1 = constant a.e. on $\{x \in [a,\ b-\lambda] : w_2(x) > 0\}$ or $\{x \in [b-\lambda,\ b] : w_2(x) < 1\}$. All signs of inequalities in (3.3.1) are reverse if w_1 is non-increasing.

Now we present the following results given by Gajek and Okolewski (1997).

Theorem 3.3.1.

Let $n \ge 0$ and $k \ge 2$. The for $\alpha = 2i-1, i \in N$, we have

$$\text{(i) } E\left[R_{n,k}^{\alpha}\right] \geq \begin{cases} q(n,k)\int_0^{1/q(n,k)}\left[F^{-1}(x)\right]^{\alpha} dx, & \text{if } n \leq [k/2], \\ q([k/2],k)\int_0^{1/q([k/2],k)}\left[F^{-1}(x)\right]^{\alpha} dx, & \text{if } n > [k/2]. \end{cases}$$

$$\text{(ii) } E\left[R_{n,k}^{\alpha}\right] \leq \begin{cases} q(n,k)\int_{1-1/q(n,k)}^{1}\left[F^{-1}(x)\right]^{\alpha} dx & \text{if } n > [k/2], \\ q([k/2],k)\int_{1/q([k/2],k)}^{1}\left[F^{-1}(x)\right]^{\alpha} dx, & \text{if } n \leq [k/2]. \end{cases}$$

Also if $\alpha = 2i, i \in N$, then

$$\text{(iii) } E\left[R_{n,k}^{\alpha}\right] \leq q(n,k)\ \left\{\int_0^{\lambda_1}\left[F^{-1}(x)\right]^{\alpha} dx + \int_{1-\lambda_2}^{1}\left[F^{-1}(x)\right]^{\alpha} dx\right\}$$

$$\text{(iv) } E\left[R_{n,k}^{\alpha}\right] \geq q(n,k)\int_{x_0-\lambda_1}^{x_0+\lambda_2}\left[F^{-1}(x)\right]^{\alpha} dx\,,$$

where $x_0 = F(0),\ q(n,k) = (k/n!)\left[kn/e(k-1)\right]^n$,

$$\lambda_1 = \begin{cases} r\left(n+1,-k\ln(1-x_0)\right)/q(n,k), & \text{if } x_0 \in [0,1), \\ 1/q(n,k), & \text{if } x_0 = 1, \end{cases}$$

$$\lambda_2 = \frac{1}{q(n,k)} - \lambda_1\,,$$

and

$$r(m,z) = \frac{\int_0^z x^{m-1}e^{-x}dx}{\int_0^{\infty} x^{m-1}e^{-x}dx},\ m > 0,$$

is the normalized incomplete gamma function.

Proof

Let us first write the expected value of the αth moment of the kth record statistics as follows:

$$E\left[R_{n,k}\right]^{\alpha} = \int_0^1 \left[F^{-1}(x)\right]^{\alpha} \frac{k}{n!}\left[-k\ln(1-x)\right]^{n}(1-x)^{k-1}\,dx,$$

for $\alpha \in N, k \geq 2$ and $n \geq 0$. Let us define

$$h(x,n,k) \leq \begin{cases} k(n!)^{-1}\left[-k\ln(1-x)\right]^{n}(1-x)^{k-1}, & if\ x \in [0,\ 1), \\ 0, & if\ x = 1, \end{cases}$$

and

$$q(n,k) = \max\{h(x,n,k) : x \in [0,\ 1]\}.$$

One can easily show that $q(n,k) = (k/n!)\left[nk/e(k-1)\right]^{n}$. From Lemma 3.3.1 with setting $w_1(x) = \left[F^{-1}(x)\right]^{\alpha}$ and $w_2(x) = h(x,n,k)/q(n,k)$, $x \in [0,\ 1]$, we obtain

$$E\left[R_{n,k}\right]^{\alpha} \geq q(n,k)\int_0^{\lambda} \left[F^{-1}(x)\right]^{\alpha} dx,$$

and

$$E\left[R_{n,k}\right]^{\alpha} \leq q(n,k)\int_{1-\lambda}^{1} \left[F^{-1}(x)\right]^{\alpha} dx,$$

where

$$\lambda = \frac{1}{q(n,k)}\int_0^1 w_2(x)\,dx = \frac{1}{q(n,k)}.$$

Applying Lemma 3.3.2 below, we obtain (i) and (ii). To show (iii) and (iv), we rewrite the expectation of αth moment of the kth RS's as

$$\begin{aligned} E\left[R_{n,k}\right]^{\alpha} &= \int_0^{x_0} \left[F^{-1}(x)\right]^{\alpha} \frac{k}{n!}\left[-k\ln(1-x)\right]^{n}(1-x)^{k-1}\,dx, \\ &+ \int_{x_0}^{1} \left[F^{-1}(x)\right]^{\alpha} \frac{k}{n!}\left[-k\ln(1-x)\right]^{n}(1-x)^{k-1}\,dx. \end{aligned}$$

It follows from Lemma 3.3.1

$$E\left[R_{n,k}\right]^{\alpha} \leq q(n,k)\left\{\int_0^{\lambda_1} \left[F^{-1}(x)\right]^{\alpha} dx + \int_{1-\lambda_2}^{1} \left[F^{-1}(x)\right]^{\alpha} dx\right\},$$

and

$$E[R_{n,k}]^{\alpha} \geq q(n,k)\int_{x_0-\lambda_1}^{x_0+\lambda_2}\left[F^{-1}(x)\right]^{\alpha}dx,$$

where

$$\lambda_1 = \int_0^{x_0} w_2(x)\,dx = \frac{1}{n!\,q(n,k)}\int_0^{-k\ln(1-x_0)} u^n e^{-u}du$$
$$= \frac{1}{q(n,k)}r(n+1,-k\ln(1-x_0)),$$

and

$$\lambda_2 = \int_{x_0}^{1} w_2(x)\,dx = \frac{1-\lambda_1}{q(n,k)}.$$

Lemma 3.3.2.

For $k \geq 2$ and $n \geq 0$,

$$q(n+1,k) < q(n,k) \qquad \text{for } n < [k/2],$$

and

$$q(n+1,k) > q(n,k) \qquad \text{for } n \geq [k/2].$$

Proof

The quotient $q^*(n,k) = q(n+1,k)/q(n,k)$ can be written as

$$q^*(n,k) = \left(1+\frac{1}{n}\right)^n\left[e\left(1-\frac{1}{k}\right)\right]^{-1},$$

and $q^*(0,k) = \left[e\left(1-\frac{1}{k}\right)\right]^{-1}$. Also consider

$$n_0 = \min\left\{ n \in N : \left(1+\frac{1}{n}\right)^n > e\left(1-\frac{1}{k}\right)\right\}.$$

It is easily checked that $q^*(n,k) < 1$ for $n < n_0$ and $q^*(n,k) > 1$ for $n \geq n_0$. The relations are valid directly for $k = 2$ and $k = 3$. It is enough to show that

$$\left(1+\frac{1}{[k/2]}\right)^{[k/2]} > e\left(1-\frac{1}{k}\right)$$

and

$$\left(1+\frac{1}{[k/2]-1}\right)^{[k/2]-1} > e\left(1-\frac{1}{k}\right),$$

for $k \geq 4$. In fact, the following inequalities are satisfied for all $x \geq 4$,

(i) $\left(1+\frac{2}{x}\right)^{x/2} > e\left(1-\frac{1}{x}\right),$

(ii) $\left(1+\frac{2}{x-1}\right)^{(x-1)/2} > e\left(1-\frac{1}{x}\right),$

(iii) $\left(1+\frac{1}{(x/2)-1}\right)^{(x/2)-1} < e\left(1-\frac{1}{x}\right),$

(iv) $\left(1+\frac{1}{(x-1)/2-1}\right)^{((x-1)/2)-1} < e\left(1-\frac{1}{x}\right).$

Using the following inequality

$$\left(1+\frac{1}{(x/2)-1}\right)^{(x/2)-1} < e\left(1-\frac{1}{x}\right) < \left(1+\frac{1}{(x-1)/2}\right)^{(x-1)/2-1},$$

(cf. Gajek and Okolewski (1997, Lemma 2) and observing that facts $[k/2] = k/2$, when k is even and $[k/2] = (k-1)/2$, when k is odd, we complete the proof.

Theorem 3.3.2.

For $k \geq 2$, $n \geq 0$ *and* F^{-1} *is convex on* $[0, 1/q(n,k)]$, *we have*

$$ER_{n,k} \leq \begin{cases} F^{-1}\left(\dfrac{1}{2\,q(n,k)}\right), & \text{if } n \leq [k/2], \\ F^{-1}\left(\dfrac{1}{2\,q([k/2],k)}\right), & \text{if } n > [k/2]. \end{cases}$$

If F^{-1} is concave on $[1-1/q(n,k), 1]$, *then*

$$ER_{n,k} \leq \begin{cases} F^{-1}\left(1-\dfrac{1}{2q(n,k)}\right), & \text{if } n > [k/2], \\ F^{-1}\left(1-\dfrac{1}{2q([k/2],k)}\right), & \text{if } n \leq [k/2]. \end{cases}$$

Table 3.2. Bounds of $E(R_{n,7})$ Based on Steffensen's Inequality when Compared with Grudzien and Szynal Bound

n	Theorem 3.3.1(i)	Theorem 3.3.1(ii)	Theorem 3.3.2	Bound in (3.1.5)
1	0.5176	0.9314	0.9335	1.0018
2	0.5509	0.9314	0.9335	0.9708
3	0.5585	0.9314	0.9335	0.9632
4	0.5585	0.9325	0.9345	0.9665
5	0.5585	0.9355	0.9373	0.9760
6	0.5585	0.9394	0.9409	0.9896
7	0.5585	0.9436	0.9447	1.0061
8	0.5585	0.9479	0.9489	1.0249
9	0.5585	0.9521	0.9529	1.0455
10	0.5585	0.9561	0.9567	1.0679

Table 3.2 presents Steffensen type bounds for $ER_{n,7}$ derived by Theorem 3.3.1 and Theorem 3.3.2. These bounds are also compared with ones derived by Grudzien and Szynal (1985) (See (3.1.5)). Table 3.3 compares the lower and upper bounds from Theorem 3.3.1 with the exact values of $E[R_{n,7}]^2$.

The next theorem provides moment inequalities for RS's under moment restrictions on their distribution. For more detail one refers to Gajek and Lenic (1993).

Theorem 3.3.3.

Let r, s, k, n ∈ N be such that $r \le k$ *and* $s \le n$. *Suppose* $ER_{n,k}^{r} = 0$ *and* $\alpha \ge 1$. *Then*

$$E\left[R_{n,k}\right]^{\alpha} \le a(n,k)\left\{\frac{E\left[R_{2s,2r-1}\right]^{2}}{a(2r-1,2s)}\right\}^{1/2}$$

$$\left\{\frac{E\left[R_{2n-2s,2k-2r+1}\right]^{2\alpha-2}}{a(2k-2r+1,2n-2s)}-\left[\frac{E\left[R_{n-s,k-r+1}\right]^{\alpha-1}}{a(k-r+1,n-s)}\right]^{2}\right\}^{1/2}, \quad (3.3.2)$$

where $a(n,k) = k^{n+1}/n!$.

Proof

Since $ER_{s,r} = 0$, for every $\rho \in R$,

$$E\left[R_{n,k}\right]^{\alpha} = E\left[R_{n,k}\right]^{\alpha} - \rho E\left[R_{s,r}\right]^{\alpha}$$

$$= \int_0^1 F^{-1}(x)\left[-\ln(1-x)\right]^{s}(1-x)^{r-1}$$

$$\left\{a(n,k)\left[F^{-1}(x)\right]^{\alpha-1}\left[-\ln(1-x)\right]^{n-s}(1-x)^{k-r} - \rho\, a(r,s)\right\} dx.$$

Applying the Cauchy-Schwarz inequality and minimizing with respect to the parameter *a* leads to the requested result. Further, the inequality in (3.3.2) becomes equality iff

$$F^{-1}(x)\left[-\ln(1-x)\right]^{s}(1-x)^{r-1} = c\left\{a(n,k)\left[F^{-1}(x)\right]^{\alpha-1}\right.$$

$$\left.\left[-\ln(1-x)\right]^{n-s}(1-x)^{k-r} - \rho_0\, a(r,s)\right\},$$

for some $c \ge 0$ and $x \in (0,1)$, and

$$\rho_0 = \frac{a(n,k)E\left[R_{n-s,k-r+1}\right]^{\alpha-1}}{a(r,s)a(k-r+1,n-s)}.$$

Table 3.3. Bounds of $E\left[R_{7,k}\right]^2$ Based on Steffensen's Inequality When Compared with Exact Values

n	Theorem 3.3.1(iv)	Theorem 3.3.1(iii)	Exact Value of $E\left[R_{n,7}\right]^2$
1	0.3653	0.8623	0.4070
2	0.4132	0.8706	0.5183
3	0.4223	0.8686	0.5976
4	0.4188	0.8708	0.6586
5	0.4092	0.8762	0.7075
6	0.3960	0.8834	0.7474
7	0.3806	0.8912	0.7807
8	0.3640	0.8992	0.8085
9	0.3469	0.9071	0.8321
10	0.3296	0.9145	0.8521

3.4. Error of the Mean Estimates

In this section, we present the lower and upper bounds on the error in approximating the mean of the nth value of kth record statistics, $E(R_{n,k})$, by a quantile function $Q(V(n/k))$, $k>1$, where $Q(x)=F^{-1}(x), x\in[0,1]$ and $V(x)=1-e^{-x}$. Quantile function $Q(V(n/k))$ represents the inverse of the hazard function $H(x)$. The following two theorems provide distribution free bounds based on the sign of $D(n,k)$, where

$$D(n,k)=1-F^{*}_{n,k}(V(n/k))-(1-V(n/k))\,f^{*}_{n,k}(V(n/k))$$

$$=\sum_{j=0}^{n}\frac{n^{j}e^{-n}}{j!}-k\frac{n^{n}e^{-n}}{n!}.$$

The results in this section were obtained by Raqab (2004).

Theorem 3.4.1.

(a) *For k>1, we have*

$$E(R_{n,k})-Q(V(n/k))\geq -L_1(n,k)\,\sigma \tag{3.4.1}$$

where

$$L_1(n,k)=\left[V(n/k)(1-V(n/k))\right]^{-1/2}\left[1-Pos(n;n)\right], \tag{3.4.2}$$

with $Pos(r;\mu)=\sum_{j=0}^{r}\frac{e^{-\mu}\mu^j}{j!}=P(V\le r)$; $V\sim$ *Poisson with mean* μ.

(b)For k>2, $n\ge 1$ *and* $D(n,k)\le 0$, *we have*

$$E(R_{n,k})-\psi(n/k)\le U_1(n,k)\,\sigma \tag{3.4.3}$$

where

$$U_1(n,k)=\left[V(n/k)(1-V(n/k))\right]^{-1/2} Pos(n;n). \tag{3.4.4}$$

Both the bounds (3.4.1) *and* (3.4.3) *are attained in the limit by continuous distributions converging weakly to the two-point distribution*

$$P\left(X=\mu-\sigma\left(e^{n/k}-1\right)^{-\frac{1}{2}}\right)=V(n/k)=1-P\left(X=\mu+\sigma\left(e^{n/k}-1\right)^{\frac{1}{2}}\right). \tag{3.4.5}$$

Proof

(a) The LHS of (3.4.1) can be written as

$$E(R_{n,k})-Q(V(n/k))=\int_0^1\left[F^{-1}(x)-\mu\right]\,G^*_{n,k}(dx),$$

where

$$G^*_{n,k}(x)=\begin{cases}F^*_{n,k}(x), \text{ if } x<V(n/k),\\ F^*_{n,k}(x)-1, \text{ if } x\ge V(n/k).\end{cases}$$

Function $-G^*_{n,k}(x)$ is starting from 0, concave-decreasing from 0 at 0 to $-F^*_{n,k}(V(n/k))$, at $V(n/k)$, jumping up to $1-F^*_{n,k}(V(n/k))>0$, and then concave-convex decreasing to 0 at $x=1$ with inflection point $V(n/(k-1))$. Thus, the greatest convex minorant is a two-piece broken line with knots (0,0), $(V(n/k),-F_{n,k}(V(n/k)))$, and (1, 0). Its derivative is

$$P\left(-g_{n,k}^{*}\right)(x)=\begin{cases}-\dfrac{F_{n,k}^{*}\left(V(n/k)\right)}{V(n/k)}, & \text{if } x<V(n/k),\\[2ex] \dfrac{F_{n,k}^{*}\left(V(n/k)\right)}{1-V(n/k)}, & \text{if } x\geq V(n/k).\end{cases}$$

The above function is projection of $-g_{n,k}^{*}(x)$ into the cone of nondecreasing functions. It follows from Lemma 3.1.1 that

$$Q(V(n/k))-E(R_{n,k})\leq\int_0^1\left[F^{-1}(x)-\mu\right]P\left(-g_{n,k}^{*}(x)\right)dx\leq\left\|P\left(-g_{n,k}^{*}(x)\right)\right\|\sigma,$$

where

$$\left\|P\left(-g_{n,k}^{*}(x)\right)\right\|^2=\left[V(n/k)(1-V(n/k))\right]^{-1}F_{n,k}^{*2}\left(V(n/k)\right).$$

From (3.1.2), the lower bound $L_l(n,k)$ is simplified and given in (3.4.2). The lower bound in (3.4.1) is attained for $F^{-1}(x)-\mu=c\left[P(-g_{n,k}^{*}(x))\right]$, where c is a constant. Condition $\int_0^1\left[F^{-1}(x)-\mu\right]^2dx=\sigma^2$ forces

$$F^{-1}(x)-\mu=\frac{P\left(-g_{n,k}^{*}(x)\right)}{\left\|P\left(-g_{n,k}^{*}\right)\right\|}\sigma,$$

which leads to the two-point distribution given in (3.4.5) as a limiting sequence of continuous distributions functions.

(b) Function $G_{n,k}^{*}$ is starting from 0, convex increasing to $F_{n,k}^{*}\left(V(n/k)\right)>0$, jumping down to $F_{n,k}^{*}\left(V(n/k)\right)-1<0$, and convex-concave increasing to 0 at 1 with inflection point

$$\mathrm{V}(n/(k-1))\in\left(V(n/k),1\right).$$

$$Pg_{n,k}^{*}\left(V(n/k)\right)=\frac{1-F_{n,k}^{*}\left(V(n/k)\right)}{1-V(n/k)}\leq g_{n,k}^{*}\left(V(n/k)\right)=f_{n,k}^{*}\left(V(n/k)\right),$$

or equivalently $D(n,k) \le 0$. Therefore the line segment joining $(V(n/k), F_{n,k}(V(n/k)) - 1)$ with (1, 0) entirely passes below the graph of $G^*_{n,k}$. The projection of $g^*_{n,k}$ is given by

$$Pg^*_{n,k}(y) = \begin{cases} \dfrac{F^*_{n,k}(V(n/k)) - 1}{V(n/k)}, & \text{if } x < V(n/k), \\[2ex] \dfrac{1 - F^*_{n,k}(V(n/k))}{1 - V(n/k)}, & \text{if } x \ge V(n/k). \end{cases}$$

From Lemma 3.1.1,

$$E(R_{n,k}) - Q\left(V\left(\frac{n}{k}\right)\right) \le \int_0^1 \left[F^{-1}(x) - \mu\right] Pg^*_{n,k}(x)\, dx \le \left\| Pg^*_{n,k} \right\| \sigma,$$

where

$$\left\| Pg^*_{n,k} \right\|^2 = \left[V(n/k)(1 - V(n/k))\right]^{-1} \left[1 - F^*_{n,k}(V(n/k))\right]^2.$$

Using again (3.1.2), the upper bound in (3.4.4) follows. The inequality becomes equality if

$$\frac{F^{-1}(x) - \mu}{\sigma} = \frac{Pg^*_{n,k}(x)}{\left\| Pg^*_{n,k} \right\|},$$

which leads in the limit to the two-point marginal distribution given in (3.4.5).

Theorem 3.4.2.

For (i) k = 1, 2, and $n \ge 1$ or (ii) $D(n,k) \ge 0$, we have

$$E(R_{n,k}) - Q\left(V\left(\frac{n}{k}\right)\right) \le U_2(n,k)\, \sigma, \tag{3.4.6}$$

where

$$U_2^2(n,k)= \quad [V(n/k)]^{-1}[Pos(n;n)]^2 + k\left(\frac{k}{2k-1}\right)^{2n+1}\binom{2n}{n}$$
$$\left[Pos\left(2n;\frac{n(2k-1)}{k}\right)-Pos\left(2n;(2k-1)\alpha^*\right)\right], \qquad (3.4.7)$$

with $\alpha^* = -\ln(1-\alpha)$ *and* $\alpha \in (V(n/k), V(n/(k-1)))$ *being the unique solution to*

$$\frac{1-\alpha}{k} f_{0,k}^*(\alpha) + \frac{1-\alpha}{k}\sum_{j=1}^{n-1} f_{j,k}^*(\alpha) - \frac{(k-1)}{k}(1-\alpha) f_{n,k}^*(\alpha) = 0. \qquad (3.4.8)$$

The upper bound is attained in the limit by continuous distributions functions converging weakly to cdf F (x) of the form

$$F(x)=\begin{cases} 0, & \text{if } x<\mu+\sigma\dfrac{F_{n,k}^*(V(n/k))-1}{V(n/k)\,U_2(n,k)}, \\[2ex] V(n/k), & \text{if } \mu+\sigma\dfrac{F_{n,k}^*(V(n/k))-1}{V(n/k)U_2(n,k)} \le x < \mu+\sigma\dfrac{f_{n,k}^*(V(n/k))}{U_2(n,k)} \\[2ex] f_{n,k}^{*-1}\left(U_2(n,k)\dfrac{x-\mu}{\sigma}\right), & \text{if } \mu+\sigma\dfrac{f_{n,k}^*(V(n/k))}{U_2(n,k)} \le x < \mu+\sigma\dfrac{f_{n,k}^*(\alpha)}{U_2(n,k)}, \\[2ex] 1, & \text{if } x \ge \mu+\sigma\dfrac{f_{n,k}^*(\alpha)}{U_2(n,k)} \end{cases}$$

(3.4.9)

Proof

On noting that $D(n,k)=a(n)-k\,b(n)$ where $a(n)=P(V\le n)\in(0,1)$ and $b(n)=P(V=n)$, we conclude that $D(n,k)\ge 0$ for both cases (i) and (ii). In this case, the greatest convex minorant has the following form:

$$\overline{G}^*_{n,k}(x) = \begin{cases} \dfrac{F^*_{n,k}(V(n/k))-1}{V(n/k)}x, & \text{if } x < V(n/k), \\ F^*_{n,k}(x)-1, & \text{if } V(n/k) \le x < \alpha, \\ f^*_{n,k}(\alpha)(x-1), & \text{if } x \ge \alpha, \end{cases}$$

where α is determined by solving the equation

$$1 - F^*_{n,k}(\alpha) = f^*_{n,k}(\alpha)(1-\alpha),\ \alpha \in (V(n/k), V(n/(k-1))).$$

The above equation is rewritten in terms of densities of kth record values as appeared in (3.4.8). The uniqueness of α follows from the fact that the linear combination of the densities of kth records does not exceed the number of sign changes of the coefficients (see, e.g., Gajek and Okolewski (2003), and Raqab (2004)).

From the projection inequality and the Schwarz inequality, we have

$$E(R_{n,k}) - Q\left(V\left(\frac{n}{k}\right)\right) \le \int_0^1 \left[F^{-1}(x)-\mu\right] Pg^*_{n,k}(x)dx \le \left\|Pg^*_{n,k}\right\| \sigma,$$

where

$$\left\|Pg^*_{n,k}\right\|^2 = [V(n/k)]^{-1}\left[1 - \mathrm{F}^*_{\mathrm{n,k}}(V(n/k))\right]^2 + \int_{V(n/k)}^{\alpha} f^{*2}_{n,k}(x)\,dx + f^{*2}_{n,k}(\alpha)(1-\alpha),$$

which is simplified and rewritten in the form of (3.4.7). The inequality in (3.4.6) becomes equality if

$$F^{-1}(x) - \mu = \frac{Pg^*_{n,k}(x)}{\left\|Pg^*_{n,k}\right\|}\sigma,$$

which leads to (3.4.9). The distribution function in (3.4.9) has two atoms of measures $V(n/k)$ and $1-\alpha$, respectively at the ends of support intervals.

In the special case $n = 1, k \ge 3, \alpha = V(1/k(k-1))$ and then the bound in (3.4.7) reduces to sharp bound

$$U_2^2(1,k) = 4[V(1/k)]^{-1} e^{-2} + 2k\left(\frac{k}{2k-1}\right)^3 \left[Pos\left(2; \frac{2k-1}{k}\right) - Pos\left(2; \frac{2k-1}{k(k-1)}\right)\right].$$

Theorem 3.4.3.

For k = 1 (ordinary record statistics), we have

$$-L_3(n)\sigma \le E(R_{n,1}) - F^{-1}(V(n)) \le U_3(n)\sigma$$

where

$$L_3(n) = [V(n)(1-V(n))]^{-1/2}[1 - Pos(n;n)],$$

and

$$U_3^2(n) = [V(n)]^{-1}[Pos(n;n)]^2 + \binom{2n}{n} Pos(2n;n).$$

The lower and upper bounds are attained in the limit by continuous distributions functions converging to the following forms of distributions:

$$P\left(X = \mu - \sigma(e^n - 1)^{-1/2}\right) = V(n) = 1 - P\left(X = \mu + \sigma(e^n - 1)^{1/2}\right)$$

and

$$F(x) = \begin{cases} 0, & \text{if } x < \mu + \sigma \dfrac{F^*_{n,1}(V(n)) - 1}{V(n)U_3(n)}, \\ V(n), & \text{if } \mu + \sigma \dfrac{F^*_{n,1}(V(n)) - 1}{V(n)U_3(n)} \le x < \mu + \sigma \dfrac{f^*_{n,1}(V(n))}{U_3(n)}, \\ f^{*-1}_{n,1}\left(U_3(n)\dfrac{x-\mu}{\sigma}\right), & \text{if } x \ge \mu + \sigma \dfrac{f^*_{n,1}(V(n))}{U_3(n)}, \end{cases}$$

Proof

Function $G^*_{n,1}(x)$ is starting from 0, convex-increasing on $[0, V(n))$, and negative increasing to 0 on $[V(n), 1]$. Thus the greatest convex minorant $\overline{G}^*_{n,1}$ of $G^*_{n,1}$ has the form:

$$\overline{G}^*_{n,1}(x) = \begin{cases} \dfrac{F^*_{n,1}(V(n)) - 1}{V(n)} x, & \text{if } x < V(n), \\ F^*_{n,1}(x) - 1, & \text{if } x \ge V(n). \end{cases}$$

In this case the greatest convex minorant is a two-piece broken line with knots (0, 0), $(V(n), -F_{n,1}(V(n)))$, and (1, 0). The resulting bounds and the corresponding probability distributions given in Theorem 3.4.3 can be concluded based on the arguments similar to those in proof of Theorem 3.4.1 and Theorem 3.4.2.

If X has continuous cdf F, the hazard function $H(X)$ has a standard exponential distribution $Exp(1)$. It follows that, $R_{n,k} \overset{d}{=} Q(V(R^*_{n,k}))$, where $R^*_{n,k}$ is the nth value of the kth records of $Exp(1)$ sequence. The strong law of large numbers assures that $R^*_{n,k} / n \xrightarrow{a.s.} 1 \text{ as } n \to \infty$. Also, if,

$$\lim_{t\to\infty} \frac{Q(V(t + x\sqrt{t})) - Q(V(t))}{Q(V(t+\sqrt{t})) - Q(V(t))} = x, \ \forall x, \tag{3.4.10}$$

then

$$\frac{R_{n,k} - Q\left(V\left(\frac{n}{k}\right)\right)}{Q\left(V\left(\frac{n}{k} + \sqrt{\frac{n}{k}}\right)\right) - Q\left(V\left(\frac{n}{k}\right)\right)} \xrightarrow{d} N(0,1). \tag{3.4.11}$$

(cf. Resnick, 1973). The condition in (3.4.10) holds in various distributions. For example, Weibull, logistic and extreme records are asymptotically normal. A consequence of that, $R_{n,k} \overset{d}{\approx} Q(V(nZ/k))$ (in which $\overset{d}{\approx}$ denotes "approximately distributed") where $Z \overset{d}{=} N$ (1, 1).

The bounds on the error in approximating the mean of nth value of kth records $R_{n,k}$ by a quantile $Q(V(n/k))$ is presented by combining the Moriguti's approximation with the Schwarz inequality. These bounds can also be improved by using the Hölder's inequality, instead of the Schwarz one. In this case, the resulting error evaluations can be measured in terms of scale units generated by absolute central population moments for the expectations of k-records.

Direct computations show that if *(i)* $k = 2$, $n \geq 1$ or *(ii)* $k = 3$, $n \geq 4$ or *(iii)* $k = 4$, $n \geq 7$, then $D(n,k) > 0$. While if *(i)* $n = 1$, $k \geq 3$ or *(ii)* $n = 1, 2; k \geq 4$ or *(iii)* $n = 1, 2, 3$; $k \geq 5$, then

$D(n,k) < 0$. When $n \leq k, k > 1$, the bounds established in this section represent the lower and upper bounds on the bias in estimating a population quantile $Q(V(p)) = F^{-1}(V(p))$, where $p \simeq n/k$ via a record statistic $R_{n,k}$. Further, the difference between the bounds in (3.4.1) and (3.4.3) is $\phi(p) = e^{p/2}(V(p))^{-1/2}$. Function $\phi(p)$ is decreasing to 2 at $p_0 = \ln 2$ and then increasing to $e(e-1)^{-1/2}$ at $p = 1$. Hence we can roughly conclude that the error decreases as n increases up to $n = 0.6\ k$ and then increases. This shows that the minimized error occurs when n is roughly 0.6 k. The approximation of $E(R_{n,k})$ by expanding $R_{n,k} = Q\left(V\left(R^*_{n,k}\right)\right)$ in a Taylor series around the point $Q(V(n/k))$ is expected to work quite well when n is large. With Maple or Mathematica procedures, one can compute series approximations for the means, variances and covariances of kth record statistics up to any desired order.

Chapter 4

Bounds for Some Restricted Families

Bounds for expectations of record values from general i.i.d samples are presented here and are obtained for symmetric unimodal, decreasing failure density and decreasing failure rate distributions. Results of Section 4.1 come from Raqab and Rychlik (2002). Section 4.2 and 4.3 are based on Gajek and Okolewski (2003). Bounds from the symmetric populations with general n and k are still not established in a rigorous manner. Raqab (1997) obtained numerical bounds from symmetric parents. The main difficulty was lying in describing the variability of the function obtained by folding the pdf of nth value of kth records of the uniform sample about ½. Raqab and Rychlik (2004) established bounds for the 2nd records from symmetric parent distributions. The trivial case $n = 0$ is excluded from the results below. If the parent distribution F is symmetric, then the folded function $\delta_{o,k}(x) = f^*_{0,k}(x) - f^*_{0,k}(1-x)$, $x \in [1/2, 1]$ is a non decreasing and convex function. Therefore the projection function of $\delta_{0,k}(x)$ is $P\delta_{0,k}(x) = \delta_{0,k}(x)$ and the resulting bound can be obtained directly. If the families of life distributions are the ones with decreasing failure probability and rate, then

$$ER_{0,k} = E(X_{1,k}) \leq \mu \leq \mu_2^{1/2},$$

where μ_2 is the second moment of F. In fact, for $n = 0$, the density $f_{0,k}(x)$ is nondecreasing concave function and its projection is constant.

4.1. Bounds for Second Records

Here we develop sharp bounds for the 2nd records with $k = 2$, which is the most natural replacement for the standard first records. If F is symmetric about μ, we can write

$$ER_{n,k} - \mu = \int_{1/2}^{1} \left[F^{-1}(x) - \mu\right] \delta_{n,k}(x)\, dx, \tag{4.1.1}$$

where

$$\delta_{n,k}(x)=f^*_{n,k}(x)-f^*_{n,k}(1-x), \tag{4.1.2}$$

and

$$\Delta_{n,k}(x)=\int_{1/2}^{x}\delta_{n,k}(u)\,du, \qquad 1/2\le x\le 1,$$

stands for its antiderivative. Applying the Cauchy-Schwarz inequality, we obtain the following rough bound

$$B^S_C(n,k)=\frac{k^{n+1}}{\sqrt{2}}\left\{\frac{1}{(2k-1)^{2n+1}}\binom{2n}{n}-\mathrm{A}_{n,k}\right\}^{1/2}, \tag{4.1.3}$$

where

$$\mathrm{A}_{n,k}=\frac{1}{(n!)^2}\int_0^1[\ln x\,\ln(1-x)]^n[x(1-x)]^{k-1}\,dx.$$

In the special case where $k=1, A_{n,k}$ reduces to A_n defined in (2.1.6). The above bound in (4.1.3) is nonsharp bound except for the case of ordinary records (k=1) (see Nagaraja, 1978). Applying directly the Hölder inequality to (4.1.1), we obtain

$$\begin{aligned} ER_{n,k}-\mu &\le \left\|F^{-1}-\mu\right\|_p\left\|\delta_{n,2}\right\|_q \\ &= B^S_H(n,p)\sigma_p, \end{aligned} \tag{4.1.4}$$

where $q=+\infty, p/(p-1)$, and 1 for p = 1, 1< p < $+\infty$, and $p=+\infty$, respectively. Therefore the bounds in (4.1.4) have the following form:

$$B^s_H(n,k,p)=\begin{cases}\dfrac{1}{2}\sup\limits_{x\in(1/2,1)}\left|\delta_{n,k}(x)\right|, & if\ p=1,\\ \dfrac{1}{2^{1/p}}\left[\int_{1/2}^{1}\left|\delta_{n,k}(x)\right|^q dx\right]^{1/q}, & if\ 1\le p<+\infty,\\ \int_{1/2}^{1}\left|\delta_{n,k}(x)\right|dx, & if\ p=+\infty.\end{cases} \tag{4.1.5}$$

If 1 < p<+∞. The equality in (4.1.4) holds iff

$$F^{-1}(x)-\mu=\alpha\, g_p\left(\delta_{n,k}(x)\right), \tag{4.1.6}$$

for $\alpha>0$ and the monotone function $g_p(x)=|x|^{q/p}$ sgn (x). Note that $F^{-1}(x)-\mu$ is necessarily nondecreasing and nonnegative on [1/2, 1]. For k = 1, function (4.1.2) is the difference of increasing and decreasing functions which coincide at ½. Therefore the right-hand side of (4.1.6) is strictly increasing and positive on (½, 1), and the bound (4.1.3) is attained by the distribution symmetric about μ and defined by

$$F(x)=\delta_{n,1}^{-1}\left(g_p^{-1}\left(2^{1/p}\left\|g_p o\delta_{n,1}\right\|_p \frac{x-\mu}{\sigma_p}\right)\right),$$

on the right to the symmetry center μ (see Nagaraja (1978), and Rychlik (2001, p. 135)).

Generally for $k\geq 2$, bounds (4.1.5) are nonsharp bounds. Projections of $\delta_{n,k}$ onto the family of nondecreasing functions in the Hilbert space $L^p([1/2,1],\, dx)$ is needed to optimize the bounds in (4.1.5). The following auxiliary lemma describes the variability of $\delta_{n,2}(x)$ and will be helpful in determining the projection of $\delta_{n,2}(x)$ onto the family of nondecreasing functions in $L^2([1/2,\,1],\, dx)$.

Lemma 4.1.1.

For every n = 1, 2, ..., there exists a unique $x_n\in\left[1-e^{-n},\,1\right)$ *such that* $\delta_{n,2}(x)$ *is increasing on* $[1/2,\, x_n)$ *and decreasing on* $(x_n,\, 1]$, *with* $\delta_{n,2}(1/2)=\delta_{n,2}(1)=0$.

Proof

The derivative of $f_{n,2}^*(x)$ can be written as

$$f_{n,2}^{*'}(x)=\frac{2^{n+1}}{n!}\tau(u(x)),$$

where

$$\tau(x)=nx^{n-1}-x^n,\qquad 0<x<+\infty,$$

$$u(x)=-ln(1-x),\qquad 0<x<1.$$

Furthermore, $\delta_{n,2}(x)$ vanishes at the ends of the interval $[1/2, 1]$ with

$$\delta'_{n,2}(x) = f^{*'}_{n,2}(x) + f^{*'}_{n,2}(1-x), \quad 1/2 \le x \le 1.$$

For the special case $n = 1, \delta'_{1,2}(x) = 4[2 + \ln x + \ln(1-x)]$. Since

$$\delta''_{1,2}(x) = \frac{4(1-2x)}{x(1-x)} < 0, \ 1/2 < x < 1,$$

we observe that the first derivative $\delta'_{1,2}(x)$ strictly decreases from $\delta'_{1,2}(1/2) = 8(1 - \ln 2) >$ 0 to - ∞ at 1, and has a unique zero at $x_1 = 0.5 + (0.25 - e^{-2})^{1/2} \in (1 - e^{-1}, 1)$.

Function $\tau(x)$ starts from 0, increases on $[0, n-1]$, decreases on $[n-1, +\infty)$ to $-\infty$ and has a zero at $x = n$, whereas $u(x)$ is strictly increasing from 0 to $+\infty$. Consequently, $f^{*'}_{n,2}(x)$ increases on $[0, 1 - e^{-(n-1)}] \supset [0, 1/2]$ from 0 at 0 to a positive maximum $f^{*'}_{n,2}(1 - e^{-(n-1)})$, and then decreases on $[1 - e^{-(n-1)}, 1] \subset (1/2, 1]$ to $-\infty$ at 1, and crosses the horizontal axis at $1 - e^{-n} \in (1/2, 1)$. Since $f^{*'}_{n,2}(1-x) > 0, 1/2 < x < 1,$ we have $\delta'_{n,2}(x) > f^{*'}_{n,2}(x)$ there, and, in consequence, $\delta'_{n,2}(x) > 0$ for $1/2 < x < 1 - e^{-n}$. Since $f^{*'}_{n,2}(1-x), 1/2 < x < 1,$ is decreasing, $\delta'_{n,2}(x)$ is also decreasing on $[1 - e^{-(n-1)}, 1]$. In particular, it decreases from $\delta'_{n,2}(1 - e^{-n}) > 0$ to $\delta'_{n,2}(1) = -\infty$, and has a unique zero at some $x_n \in (1 - e^{-n}, 1)$. Therefore function $\delta_{n,2}(x)$ is increasing on $[1/2, x_n)$ and decreasing on $(x_n, 1]$, which completes the proof.

Lemma 4.1.2.

For given $n \ge 1$, there exists a unique $y_n \in (1/2, x_n)$ defined as the solution to equation

$$\delta_{n,2}(y_n)(1 - y_n) = \int_{y_n}^{1} \delta_{n,2}(x)dx, \tag{4.1.7}$$

such that for

$$P\delta_{n,2}(x)=\delta_{n,2}(\min\{x,y_n\}),\quad 1/2\le x<1, \tag{4.1.8}$$

and every non-decreasing $w\in L^1([1/2,1],\,dx)$, *we have*

$$\int_{1/2}^{1} w(x)\,\delta_{n,2}(x)\,dx\le\int_{1/2}^{1} w(x)P\delta_{n,2}(x)\,dx \tag{4.1.9}$$

with the equality iff

$$w(x)=const,\quad y_n<x<1. \tag{4.1.10}$$

Proof

We show that $P\delta_{n,2}(x)$ is the derivative of the greatest convex minorant $\overline{\Delta}_{n,2}(x)$ of the anti-derivative $\Delta_{n,2}(x)$ of $\delta_{n,2}(x)$. From Lemma 4.1.1, the function $\Delta_{n,2}$ is convex increasing on $[1/2,\,x_n)$, and concave increasing on $(x_n,\,1]$. Therefore the greatest convex minorant of $\Delta_{n,2}(x)$ has the form

$$\overline{\Delta}_{n,2}(x)=\begin{cases}\Delta_{n,2}(x), & \text{if } 1/2\le x\le y_n,\\ \delta_{n,2}(y_n)(x-1)+\Delta_{n,2}(1), & \text{if } y_n\le x\le 1,\end{cases}$$

for a unique $y_n\in(1/2,x_n)$ satisfying (4.1.7). Differentiating the function, we conclude that (4.1.8) is the projection of $\delta_{n,2}$ onto the convex cone of nondecreasing functions in $L^2([1/2,1],\,dx)$ (cf. Rychlik, 2001, pp. 14-16). Applying the Hölder inequality to (4.1.1) and using (4.1.9), we obtain

$$\begin{aligned}E(R_{n,2}-\mu)&\le\int_{1/2}^{1}\left[F^{-1}(x)-\mu\right]P\delta_{n,2}(x)\,dx\\&\le B_{PH}^{S}(n,p)\sigma_p,\end{aligned} \tag{4.1.11}$$

where

$$\begin{aligned}B_{PH}^{S}(n,p)&=2^{-1/p}\left\|P\delta_{n,2}\right\|_q\\&=2^{-1/p}\left\{\int_{1/2}^{y_n}\left[\delta_{n,2}(x)\right]^q dx+\left[\delta_{n,2}(y_n)\right]^q(1-y_n)\right\}^{1/q}.\end{aligned} \tag{4.1.12}$$

The latter inequality in (4.1.11) becomes equality iff

$$F^{-1}(x)-\mu=\alpha\left[P\delta_{n,2}(x)\right]^{q/p}, \quad 1/2<x<1, \tag{4.1.13}$$

for some $\alpha \geq 0$. The right-hand side is nonnegative and nondecreasing and it is constant on $(y_n, 1)$, by (4.1.10). Further this condition is necessary and sufficient for the equality in the former inequality of (4.1.11). Condition $\left\|F^{-1}-\mu\right\|_p^p=\sigma_p^p/2$ forces that constant

$$\alpha=\frac{\sigma_p}{2^{1/p}\left\|P\delta_{n,2}\right\|_q^{q/p}}=\frac{\alpha_p}{\left[2B_{PH}^S(n,p)\right]^{q/p}}$$

is uniquely determined. As a consequence of that a general form of cdf *F(x)* can be formalized. The optimal bound and the respective *cdf F(x)* where the bound is attained are summarized in the following theorem:

Theorem 4.1.1.

For a symmetric parent distribution with finite mean μ *and* p^{th} *absolute central moment* σ_p^p, $1<p<+\infty$, *and arbitrary* $n\geq 1$, *we have*

$$\frac{ER_{n,2}-\mu}{\sigma_p}\leq B_{PH}^S(n,p),$$

where the right-hand side is defined in (4.1.12) *with unique* $y_n \in (1/2,\, x_n)$ *solving equation* (4.1.7). *The bound is attained in the limit by continuous distributions functions converging weakly to the cdf's of the form*

$$F(x)=\begin{cases}0, & \text{if } x\leq\alpha_1,\\ \delta_{n,2}^{-1}\left(2B_{PH}^S(n,p)\left|\dfrac{x-\mu}{\sigma_p}\right|^{p/q}\operatorname{sgn}(x-\mu)\right), & \text{if } \alpha_1\leq x<\alpha_2,\\ 1, & \text{if } x\geq\alpha_2,\end{cases} \tag{4.1.14}$$

where

$$\alpha_1=\mu-\sigma_p\left[\frac{\delta_{n,2}(y_n)}{2B_{PH}^S(n,p)}\right]^{q/p},\quad \alpha_2=\mu+\sigma_p\left[\frac{\delta_{n,2}(y_n)}{2B_{PH}^S(n,p)}\right]^{q/p}.$$

Distribution function (4.1.14) has the symmetry center μ, a smooth component on finite support [α_1, α_2] and two atoms of measure $(1-y_n)$ at the ends of the support interval. The following theorems provide the optimal bounds for the extreme cases $p = 1$ and $p = \infty$.

Thereom 4.1.2.

Let a symmetric population distribution have mean μ and first central absolute moment $\sigma_1 < +\infty$. Then for every $n \geq 1$,

$$\frac{ER_{n,2}-\mu}{\sigma_1} \leq B_{PH}^{S}(n,p) = \frac{\delta_{n,2}(y_n)}{2}, \qquad (4.1.15)$$

where y_n is uniquely defined in (4.1.7). The bound becomes equality for continuous distributions converging weakly to a sequence of three-point distribution

$$P\left(X_1 = \mu \pm \frac{\sigma_1}{2(1-y_n)}\right) = 1 - y_n,$$

$$P(X_1 = \mu) = 2y_n - 1$$

Proof

We have

$$\begin{aligned} ER_{n,2} - \mu &\leq \int_{1/2}^{1} \left[F^{-1}(x) - \mu\right] P\delta_{n,2}(x)\, dx \\ &\leq \sup_{x \in (1/2,1)} P\delta_{n,2}(x) \int_{1/2}^{1} \left[F^{-1}(x) - \mu\right] dx \\ &= \frac{\delta_{n,2}(y_n)}{2}\sigma_1 . \end{aligned}$$

The former and latter bounds are attained iff $F^{-1}(x) - \mu = const > 0$ for $y_n < x < 1$, and 0 for $1/2 < x < y_n$, respectively. Using the fact that $F^{-1} - \mu$ is a symmetric function about 1/2, we define a three-point distribution symmetric about μ with weights $1 - y_n$ of the extreme atoms.

Theorem 4.1.3.

For a population symmetrically distributed on a finite interval $[\mu-\sigma_\infty, \mu+\sigma_\infty]$, *and arbitrary* $n \geq 1$, *we have*

$$\frac{ER_{n,2}-\mu}{\sigma_\infty} \leq B_{PH}^{S}(n,\infty) = 2\Gamma(n+1,\ 2\ln 2)-1, \tag{4.1.16}$$

where

$$\Gamma(b,x) = \frac{1}{\Gamma(b)}\int_x^\infty t^{b-1}e^{-t}\,dt,$$

for positive b and x. The bound in (4.1.16) is attained by the symmetric two-point distribution supported on points $\mu \pm \sigma_\infty$.

Proof.

Here,

$$\begin{aligned} ER_{n,2}-\mu &\leq \int_{1/2}^{1}\left[F^{-1}(x)-\mu\right]P\delta_{n,2}(x)\,dx \\ &\leq \sup_{x\in(1/2,1)}\left(F^{-1}(x)-\mu\right)\int_{1/2}^{1}P\delta_{n,2}(x)dx \\ &= \overline{\Delta}_{n,2}(1)\sigma_\infty. \end{aligned} \tag{4.1.17}$$

By definition,

$$\overline{\Delta}_{n,2}(1) = \Delta_{n,2}(1) = \int_{/2}^{1}\left[f_{n,2}^{*}(x)-f_{n,2}^{*}(1-x)\right]dx.$$

The latter inequality of (4.1.17) becomes equality iff $F^{-1}(x)-\mu=\sigma_\infty$ on (1/2, 1), which satisfies the former equality. The condition, together with the symmetry property, uniquely determines the two-point distribution for which the equality holds in the limit.

To compute the sharp bounds on the upper deviations of the expectations of the nth values of the second record statistics $ER_{n,2}$, from the population mean μ, we need to determine numbers $y_n(n \geq 1)$ by solving equation (4.1.7), whose right-hand side can be simplified and rewritten in terms of Poisson sum of probabilities. The simplified equivalent form of (4.1.7) is given by

$$\delta_{n,2}(y_n)(1-y_n)=\sum_{j=0}^{n}\frac{[-2\ln(1-y_n)]^j(1-y_n)^2+[-2\ln y_n]^j y_n^2}{j!}-1$$

Using (4.1.12), we evaluate the sharp bounds $B_{PH}^S(n,p)$ for $p = 2,3$. For $p = 1$ and $+\infty$, we apply (4.1.15) and (4.1.16), respectively. These bounds are compared with rough ones $B_H^S(n,p)$, obtained by means of the Hölder inequality and described in (4.1.5) for $p = 1,2,3$ and ∞. The results are presented in Table 4.1 for $1 \le n \le 10$. We observe that the bounds decrease as p increases. The relative discrepancy between $B_H^S(n,p)$, and $B_{PH}^S(n,p)$ decreases to zero with increase of parameter p. Decrease of discrepancy occurs as well, when we fix p and increase n. For $p=\infty$, it is easily checked that $B_H^S(n,\infty)=B_{PH}^S(n,\infty)$.

Table 4.1. Bounds on the Expectations of Second Records in Various Scale Units

	$p = 1$		$p = 2$		$p = 3$		$p = \infty$
n	$B_H^S(n,1)$	$B_{PH}^S(n,1)$	$B_H^S(n,2)$	$B_{PH}^S(n,2)$	$B_H^S(n,3)$	$B_{PH}^S(n,3)$	$B_{PH}^S(n,\infty)$
1	0.2935	0.2447	0.2124	0.2070	0.2040	0.2012	0.1931
2	1.0502	0.9153	0.7460	0.7347	0.7145	0.7087	0.6736
3	1.7922	1.6240	1.0608	1.0537	0.9872	0.9839	0.8956
4	3.1259	2.8843	1.3475	1.3426	1.1752	1.1731	0.9726
5	5.6150	5.2443	1.7068	1.7029	1.3685	1.3671	0.9939
6	10.2799	9.6862	2.1789	2.1758	1.5962	1.5953	0.9988
7	19.0724	18.0920	2.7996	2.7969	1.8690	1.8683	0.9998
8	35.7342	34.0777	3.6142	3.6119	2.1953	2.1947	1.0000
9	67.4589	64.6091	4.6832	4.6811	2.5848	2.5844	1.0000
10	128.1130	123.1400	6.0861	6.0843	3.0493	3.0489	1.0000

4.2. Distributions with Decreasing Failure Probability

We now present the bounds on $ER_{n,k}$ for i.i.d. sequences with the decreasing failure probability (DFP). That is, we need to project the density $f_{n,k}^*(x)$ onto the cone of nondecreasing functions and convex. The respective quantile functions form a convex cone

$$C_1=\{Q:[0,1)\to[0,\infty):Q(0)=0,\ Q \text{ is nondecreasing and convex}\},$$

where $Q(x)$ is the quantile function of F. In order to present the main results in this section, we present the following variation diminishing property (VDP) of the record densities established by Gajek and Okolowski (2002). The result is a counterpart of the well known Schoenberg (1959) of Berstein polynomials and on the similar lines of Karlin and Studden (1966, corollary 4.1).

Lemma 4.2.1.

For every k and n, the number of zeros in (0, 1) of a linear combination $\sum_{i=0}^{\infty} a_i f_{i,k}^*(x)$ *does not exceed the number of sign changes of the sequence* $a_0,\dots,a_n$. *The first and the last sign of the sum are identical with the signs of the first and the last nonzero elements of* $a_0,\dots,a_n$ *respectively.*

Proof

Clearly

$$f_{i,k}^*\left(1-e^{-x}\right) e^{-x} = g_{i+1,k}(x) = \frac{k^{i+1}}{i!} x^i e^{-kx},\ x \geq 0.$$

This means that it is enough to show the above theorem based on the density functions $g_{i+1,k}(x)$, which are gamma densities $G(i+1,k)$. From Lemma 6 of Karlin (1956), we can easily show that for fixed *k*, the one parameter family of gamma densities is strictly totally positive of order ∞. That is; for every positive integer *i*, and arbitrary nondecreasing sequences $0 < x_1 < \dots < x_i, 0 < t_1 < \dots < t_i$, we have $\det\left[g_{t_r,k}(x_s)\right]_{r,s=1,\dots,i} > 0$. For more details about the variation diminishing property for linear combinations of densities of totally positive families, one can refer to Karlin and Studden (1966, Cor. 4.1) and Karlin (1968, Ch. 1).

Theorem 4.2.1.

Let $X_1, X_2, \dots,$ *be iid r.v.'s having distribution function F with DFP. If* $(1+1/k)^{n+1} \leq 3$, *then*

$$ER_{n,k} \leq B_{PC} = \left[1-\left(\frac{k}{k+1}\right)^{n+1}\right]\sqrt{3\mu_2}\ .$$

The equality holds iff F is the uniform distribution on $\left[0, \sqrt{3\mu_2}\right]$. *If* $k \geq 2$ *and* $(1+1/k)^{n+1} > 3$, *then*

$$ER_{n,k} \leq B_{PC} = \left\|\left(f_{n,k}^*\right)_{\min\{v_1,v_2\}}\right\| \sqrt{\mu_2}, \tag{4.2.1}$$

where v_1 *is the unique zero in* $\left(0,1-e^{-\eta}\right)$ *with*

$$\eta = \begin{cases} n-1, & \text{if } k=2, \\ \dfrac{n(2k-3)+\left[n^2+4n(k-1)(k-2)\right]^{1/2}}{2(k-1)(k-2)}, & \text{if } k \ge 3, \end{cases}$$

of the function

$$V_1 = \frac{3}{k}\sum_{i=0}^{n-1}\left[1-\left(\frac{k}{k+1}\right)^{n-i}\right] f_{i,k}^{*}(x) - k\, f_{n-2,k}^{*}(x) + \left(k-\frac{5}{2}\right) f_{n-1,k}^{*}(x), \tag{4.2.2}$$

and v_2 *is the first zero in* $\left(0,1-e^{-n/(k-1)}\right)$ *of*

$$Z_1 = \frac{1}{2k}\sum_{i=0}^{n-2}\left[3\left(\frac{k}{k+1}\right)^{n-i}-1\right] f_{i,k}^{*}(x) - \frac{(k-1)(k-2)}{4k(k+1)} f_{n-1,k}^{*}(x). \tag{4.2.3}$$

Inequality (4.2.1) becomes the equality if the quantile function of F is

$$Q(x) = \frac{\sqrt{\mu_2}}{\left\|\left(f_{n,k}^{*}\right)_{\min\{v_1,v_2\}}\right\|}\left(f_{n,k}^{*}\right)_{\min\{v_1,v_2\}}.$$

For $k=1$ *and* $n \ge 1$, $ER_{n,1} \le B_{PC} = \left[\left(\frac{2n}{n}\right)\mu_2\right]^{1/2}$ *, which becomes equality for*

$$F(x) = 1-e^{-\left(\frac{(2n)!\,x^2}{\mu_2}\right)^{1/2n}}, x>0.$$

Proof

The proof is similar to the proof of the respective result on order statistics of Gajek and Rychlik (1998). Our major goal is to determine the projection of $f_{n,k}^{*}(x)$ onto C_1. If $k = 1$ and $n \ge 1$, then $f_{n,k}^{*}(x)$ is convex increasing. Cauchy-Schwarz inequality leads to the sharp bound given in (2.1.3). If $n = 1$ and $k \ge 2$, function $f_{1,k}^{*}(x)$ is concave for $k = 2$ and concave-convex otherwise. By Lemma 1 of Gajek and Okolewski (2003), then $P f_{1,k}^{*}$ is linear.

If $n \geq 2$ and $k \geq 2$, then $f_{n,k}^{*}$ is convex-increasing on $\left(0,1-e^{-\eta}\right)$, concave increasing on $\left(1-e^{-\eta},1-e^{-n/(k-1)}\right)$ and decreasing on $\left(1-e^{-n/(k-1)},1\right)$. In this case, function $f_{n,k}^{*}$ satisfies conditions 4.29-4.33 in Rychlik (2001, p. 61). This implies that the projection of $f_{n,k}^{*}$ is $Pf_{n,k}^{*}=\left(f_{n,k}^{*}\right)_{y}, y\in(0, b)$, where b is the smallest inflection point of $f_{n,k}^{*}$. Arguments similar to those in Gajek and Rychlik (1998), we compute the corresponding values of (4.41) and (4.42) in Rychlik (2001, p. 65). The slope of linear part of this projection is

$$\alpha^{*}(y)=\frac{\int_0^1 (x-y)\left(f_{n,k}^{*}(x)-f_{n,k}^{*}(y)\right) dx}{\int_0^1 (x-y)^2 dx}.$$

Using the facts

$$f_{n,k}^{*\,'}(x)=\frac{kf_{n-1,k}^{*}(x)-(k-1)f_{n,k}^{*}(x)}{1-x}$$

and

$$\int_x^1 f_{n,k}^{*}(y)dy=\frac{1-x}{k}\sum_{i=0}^{n} f_{i,k}^{*}(x),$$

we immediately express $V_1(y)=[\alpha^{*}(y)-f_{n,k}^{*\,'}(y)](1-y),\ 0\leq y\leq 1$ in terms of sum of the densities of kth records as given in (4.2.2). Gajek and Rychlik (1998) showed that $\alpha^{*}(y)$ is the slope of the best one among the linear approximations $f_{n,k}^{*}(y)+\alpha(x-y)$ of $f_{n,k}^{*}(y)$ on $[y, 1]$. Also,

$$Z_1(y)=\frac{1}{1-y}\int_y^1\left[f_{n,k}^{*}(x)-f_{n,k}^{*}(y)-\alpha^{*}(y)(x-y)\right] dx,$$

has a simple form in (4.2.3). From Lemma 4.2.1, we observe that V_1 changes its sign once from + to -, if k = 2 and otherwise it is positive near 0 and 1, and possibly negative in a subinterval of (0,1). It is easily checked that V_1 has a single root v_1 in (0, b), where b is the smallest inflection point of $f_{n,k}^{*}$. If $(1+1/k)^{n+1}\leq 3$, $Pf_{n,k}^{*}$ is a linear function starting from the origin since Z_1 cannot have any zero in $v^{+}=\{y\in(0, b): V(y)>0\}$. If

$(1+1/k)^{n+1} > 3$, Z_1 changes its sign from – to + if k=2, and Z_1 is - + - or – otherwise. The first zero v_2 of Z_1 belongs to $\left(0, 1-e^{-n/(k-1)}\right)$. Summing up, the results follow by applying Eqs. (6) and (7) of Gajek and Okolewski (2003).

4.3. Distributions with Decreasing Failure Rate

Here we restrict ourselves to the class of all decreasing failure rate (DFR) distribution functions F which precede the exponential distribution $V(x) = 1 - e^{-x}, x > 0$. Changing the variables, leads to a representation of $ER_{n,k}$ with gamma density function given in (3.1.2). We need the projection of $g_{n,k}(x), x \geq 0$ (see (3.1.2)) onto the cone C_2, where

$C_2 = \{Q\,oV : F$ is the class of all DFR distribution functions$\}$

$= \{Q : [0,\infty) \to [0,\infty) : Q(0) = 0,\ Q$ is nondecreasing and convex$\}$,

with respect to the norm

$$\|g\|_V = \left(\int_0^\infty g^2(x)e^{-x}dx\right)^{1/2}.$$

Theorem 4.3.1.

Let $X_1, X_2, \ldots,$ be i.i.d r.v.'s having distribution function F with DFR. If $n \leq 2k-1$, then

$$ER_{n,k} \leq B_{PC}(n,k) = \frac{n+1}{k}\sqrt{\frac{\mu_2}{2}}, \tag{4.3.1}$$

which becomes the equality iff F is the exponential distribution with the scale parameter $\sqrt{\mu_2/2}$. If $k \geq 2$ and $n > 2k-1$, then

$$ER_{n,k} \leq B_{PC} = \left\|\left(g_{n,k}\right)_{\min\{v_1,v_2\}}\right\|_V \sqrt{\mu_2}, \tag{4.3.2}$$

where v_1 is the unique zero in $\left(0,\ (n-\sqrt{n})/(k-1)\right)$ of

$$V_2 = \sum_{i=0}^{n-3} \frac{n-i}{2k^2} g_{i,k}(x) - \frac{1-k^3}{k^2} g_{n-2,k}(x)$$
$$+ \frac{(k-1)^2(2k-1)}{2k^2} g_{n-1,k}(x), \tag{4.3.3}$$

and v_2 *is the first zero in* $\left(0, 1-e^{-n/(k-1)}\right)$ *of*

$$Z_2 = \sum_{i=0}^{n-2} \frac{2k-n+i}{2k^2} g_{i,k}(x) - \frac{(k-1)^2}{2k^2} g_{n-1,k}(x). \tag{4.3.4}$$

Inequality (4.3.1) *becomes the equality if the quantile function of F is*

$$Q \circ V = \frac{\sqrt{\mu_2}}{\left\| \left(g_{n,k}\right)_{\min\{v_1,v_2\}} \right\|_V} \left(g_{n,k}\right)_{\min\{v_1 v_2\}}.$$

For $k = 1$ *and* $n \geq 1$,

$$ER_{n,1} \leq B_{PC} = \left[\binom{2n}{n} \mu_2 \right]^{1/2}, \tag{4.3.5}$$

which becomes equality for

$$F(x) = 1 - e^{-\left(\frac{(2n)!\, x^2}{\mu_2}\right)^{1/2n}}, x > 0.$$

Proof

The moment bounds for k^{th} record statistics of DFR distributions can be derived based on projecting the density $g_{n,k}(x), x > 0$ onto C_2 in the norm with the exponential weight. The lines of the proof are similar to that of Gajek and Rychlik (1998). For $n = 0$, $n = 1$ with $k \geq 2$, $Pg_{n,k}$ are linear, since the functions $g_{n,k}$ are decreasing and concave-convex, respectively. For $k = 1$, $n \geq 1$, $Pg_{n,1} = g_{n,1}$. Using the series representations

$$g'_{n,k}(x) = kg_{n-1,k}(x) - (k-1)g_{n,k}(x),$$

and

$$\int_y^1 g_{n,k}(x)e^{-x}\,dx = \frac{e^{-y}}{k}\sum_{i=0}^{n} g_{i,k}(y),$$

we write analogous forms of $V_1(y)$ and $Z_1(y)$ as the following:

$$V_2(y) = \sum_{i=0}^{n-1} c_i\, g_{i,k}(y),$$

with

$$c_i = \frac{n-i}{2k^2}, i = 0,\ldots,n-3,$$

$$c_{n-2} = \frac{1-k^3}{k^2},$$

$$c_{n-1} = \frac{2k^3 - 3k^2 + 1}{2k^2},$$

and

$$Z_2(y) = e^{-y}\sum_{i=0}^{n-1} d_i\, g_{i,k}(y),$$

with

$$d_i = \frac{2k - n + i}{2k^2}, i = 0,\ldots,n-2$$

$$d_{n-1} = -\frac{(k-1)^2}{2k^2} < 0.$$

From the VDP of $g_{n,k,}$ we evaluate the zeros of V_2 and Z_2. Similar arguments to the proof of Theorem 4.2.1, we show that $V_2 < 0$ at the first inflection point b^* of $g_{n,k}(y)$. This implies that there exist $v \in (0, b^*)$ such that V_2 is + and – in $[0, v)$ and (v, b^*), respectively. If $d_0 \geq 0$, then $Pg_{n,k} = \alpha^*(0)y$ and the optimal bound simplifies to the one in (4.3.1). If

$d_0 < 0,$ then $0 > d_0 < d_1 < \ldots < d_{n-2} > 0$ and $d_{n-1} < 0.$ In this case, $Pg_{n,k} = (g_{n,k})_{\min\{v_1, v_2\}}$, where v_1 and v_2 are defined above

Table 4.2. Bounds for Expectations of kth Records from DFP and DFR Distributions

n	$U(0,\sqrt{3})$	Theorem 4.2.1	$E(1/\sqrt{2})$	Theore m 4.3.1	$B_{PC}(n,3)$	$B_C(n,3)$
0	0.4330	0.4330	0.2357	0.2357	1.0000	1.3416
1	0.7578	0.7578	0.4714	0.4714	1.0226	1.1384
2	1.0013	1.0013	0.7071	0.7071	1.1095	1.1831
3	1.1840	1.1852	0.9428	0.9428	1.2402	1.2931
4	1.3210	1.3692	1.1785	1.1785	1.4086	1.4548
5	1.4238	1.5855	1.4142	1.4142	1.6158	1.6561
6	1.5009	1.8415	1.6499	1.6707	1.8663	1.9027
7	1.5586	2.1452	1.8856	1.9738	2.1663	2.2002
8	1.6020	2.5058	2.1213	2.3323	2.5243	2.5564
9	1.6345	2.9336	2.3570	2.7567	2.9503	2.9813
10	1.6589	3.4414	2.5927	3.2596	3.4567	3.4870
11	1.6772	4.0439	2.8284	3.8560	4.0582	4.0082
12	1.6909	4.7590	3.0641	4.5636	4.7727	4.8025
13	1.7012	5.6080	3.2998	5.4036	5.6211	5.6512
14	1.7089	6.6160	3.5355	6.4012	6.6288	6.6591

Table 4.2 compares the exact values of $ER_{n,3}/\mu_2$, $n = 0,1,\ldots,14$, for the uniform $U(0,\sqrt{3})$ and the exponential distribution with scale parameter $1/\sqrt{2}$, $Exp(1/\sqrt{2})$ with the respective bounds established in Theorem 4.2.1, Theorem 4.3.1, Grudzien and Szynal (1985) ($B_C(n,k)$ given in (3.1.5)) and Raqab (1997) ($B_{PC}(n,k)$ given in (3.1.8)).

Chapter 5

SHARP BOUNDS ON RECORDS INCREMENTS

It is of interest to evaluate future records based on the observed ones. Ahsanullah (1980), Dunsmore (1983) and Awad and Raqab (2000) discussed point and interval predictions of future records of exponential distributions with location and scale parameters. In this chapter, we determine the bounds of the ordinary record increments as well as the kth record increments. Rychlik (1997) established bounds the moments on the increments from general and restricted families. Raqab (2004a) established p-norm bounds for the moments of adjacent k^{th} record increments. Danielak (2004) considered the non-adjacent increments for ordinary records and derived sharp upper bounds under general and restricted families. Danielak and Raqab (2004a, b) generalized the results of Raqab (2004a) and Danielak and Raqab (2004a, b) and presented sharp upper bounds on non-adjacent k^{th} record increments from general distributions and adjacent k^{th} record increments under restricted families, respectively.

5.1. ORDINARY ADJACENT INCREMENTS

Nagaraja (1978) derived sharp bounds for the nth value of record statistics by using Cauchy-Schwarz inequality. The bound is obtained iff $Q(u)-\mu$ and $f_n(x)-1$ is proportional, i.e. for an affine transformation of Weibul variables with the shape parameter $1/n$. Define

$$\varphi_{m,n}(x)=f_n^*(x)-f_m^*(x)$$

where $f_n^*(n)$ is the pdf of the nth ordinary record from the standard uniform distribution U(0, 1). Then

$$\frac{EI_{m,n}}{\sigma}=\frac{1}{\sigma}\int_0^1 Q(x)\,\varphi_{m,n}(x)\,dx, \tag{5.1.1}$$

where $I_{m,n} = R_n - R_m, 0 \le m < n$. For $m = n - 1$, we write $\varphi_{n-1,n}(x) = \varphi_n(x)$. For $m = 0$, the increment expectation is simply $ER_n - \mu$, which is already presented in Nagaraja (1978). Clearly (5.1.1) does not yield the sharp bound, because $\varphi_n(x)$ is not nondecreasing and so cannot be proportional to any quantile function.

The next three theorems are presented in Rychlik (1997).

Theorem 5.1.1.

For a sequence of independent identically distributed r.v.'s with a finite variance, we have for $n \ge 1$,

$$EI_{n-1,n} \le B_n \sigma, \tag{5.1.2}$$

where

$$B_n = \left\{ \binom{2n-2}{n-1} (1-y) \left[\sum_{i=0}^{2n-1} \frac{(ny)^i}{i!} + \left(1 - \frac{1}{n}\right) \frac{(ny)^{2n-1}}{(2n-1)!} \right] \right\}^{1/2}, \tag{5.1.3}$$

and $y \in (0, 1)$ *satisfies the equation*

$$-\ln(1-y) = ny. \tag{5.1.4}$$

Equality in (5.1.2) *is attained iff*

$$Q(x) = \mu + \frac{\sigma}{B_n} \varphi_n(\max\{x, y\}), \quad \mu \in R$$

Proof

Applying Cauchy-Schwarz inequality, we obtain

$$\begin{aligned} EI_{n-1,n} &\le \int_0^1 (Q(x) - \mu)\, \varphi_n(x)\, dx \\ &\le \left[\int_0^1 (Q(x) - \mu)^2\, dx \int_0^1 [P\varphi_n(x)]^2\, dx \right]^{1/2} \\ &= \left[\int_0^1 [P\varphi_n(x)]^2\, dx \right]^{1/2}. \end{aligned}$$

Also we have

$$\int_0^x f_n^*(t)\,dt = \frac{(-1)^n}{n!}\int_{1-x}^1 [\ln t]^n\,dt$$
$$= \frac{(-1)^n}{n!}\left[-(1-x)[\ln(1-x)]^n - n\int_{1-x}^1 [\ln t]^{n-1}\,dt\right]$$
$$= -(1-x)f_n^*(x) + \int_0^x f_{n-1}^*(t)\,dt,$$

and then

$$\Phi_n(x) = \int_0^x \varphi_n(t)\,dt = -(1-x)f_n^*(x). \tag{5.1.5}$$

The antiderivative of $\varphi_n(x)$, $\Phi_n(x)$ is concave decreasing, convex decreasing and convex increasing in $\left[0, 1-e^{-n+1}\right]$, $\left[1-e^{-n+1}, 1-e^{-n}\right]$ and $\left[1-e^{-n}, 1\right]$ respectively, with $\Phi_n(0) = \Phi_n(1) = 0$. It is easily observed that the greatest convex minorant $\overline{\Phi}_n$ of Φ_n is linear in [0, y] for some $y \in [1-e^{-n+1}, 1-e^{-n}]$, and coincides with Φ_n elsewhere. In fact, the value of y is to be determined based on the following equation:

$$\varphi_n(y)\,y = \Phi_n(y) \tag{5.1.6}$$

Using (5.1.5), Eq. (5.1.6) is equivalent to (5.1.4). As a consequence of that, the projection of

φ_n is $P\varphi_n(x) = \varphi_n(\max\{x, y\})$. By (5.1.5) and using

$$\int_x^1 \varphi_n(t)\,dt = (1-x)f_n^*(x),$$

we obtain

$$\int_0^1 f_n^*(t)\,dt = (1-x)\sum_{i=0}^n f_i^*(x).$$

Since

$$f_n^{*2}(x) = \binom{2n}{n}\varphi_{2n}(x) + \binom{2n-2}{n-1}f_{2n-2}^*(x),$$

we have

$$B_n^2 = \int_0^y \varphi_n^2(x)dx + \int_y^1 \varphi_n^2(x)\,dx$$

$$= \frac{\Phi_n^2(y)}{y} + \binom{2n}{n}\int_y^1 \varphi_{2n}(x)dx + \binom{2n-2}{n-1}\int_y^1 f_{2n-2}^*(x)dx$$

$$= \binom{2n}{n} f_{2n}^*(y)\frac{(1-y)^2}{y} + \binom{2n}{n} f_{2n}^*(y)(1-y) + \binom{2n-2}{n-1}(1-y)\sum_{i=0}^{2n-2} f_i^*(y)$$

$$= (1-y)\left[\binom{2n-2}{n-1}\sum_{i=0}^{2n-2} f_i^*(y) + \binom{2n}{n}\frac{f_{2n}^*(y)}{y}\right].$$

Remark 5.1.1.

A $n \to \infty$,

$$B_n = \binom{2n-2}{n-1}^{1/2}(1+o(1)) = B_C(n-1)(1+o(1)), \tag{5.1.7}$$

where $B_C(n)$ is defined in (2.1.3). That is; the extreme expected difference of nth and $(n-1)^{st}$ values of records increases at the same rate as the extreme expectation of the $(n-1)^{st}$ record. The Stirling formula yields $B_n \sim A(n-1) \sim 2^{n-1}(n\pi)^{-1/4}$. *Rychlik (1997) observed through numerical computations that the approximation* (5.1.7) *is accurate for small n.*

For symmetric distribution F, the usual way is to fold the quantile function about ½ and maximizing the integral $\int_{1/2}^1 Q(x)\delta_n(x)dx$, where $\delta_n(x) = \varphi_n(x) - \varphi_n(1-x)$. Applying Moriguti's method and Cauchy-Schwarz inequality, we have

$$EI_{n-1,n} \le \left\{\frac{1}{2}\int_{1/2}^1 [P\delta_n(x)]^2\,dx\right\}^{1/2}.$$

In practice, the life distributions with monotone failure probability and rate are of great interest. These restricted families arise in reliability theory. The problem now is to obtain the range of the expected record improvement over these classes of distributions. Further, these distributions can be defined via the convex partial ordering relations (see, e.g., van Zwet (1964)) for details. The quantile function defined in Theorem 5.1.1 is convex and the respective distribution has a decreasing failure probability. This is also a DFR distribution, since $Q(1-e^{-x})$ is convex as well. Therefore, the so obtained bounds coincide with the one in (5.1.2).

Theorem 5.1.2.

(a) For the family of life distributions with increasing failure probability (IFP) yields

$$EI_{n-1,n} \leq \frac{\sqrt{3}\,\sigma}{2^n},$$

which is attained for the uniform distribution.

(b) For the increasing failure rate (IFR) distributions

$$EI_{n-1,n} \leq \sigma,$$

which is attained for the exponential distribution.

Proof

In case (a), the cone consists of all nondecreasing concave functions integrating to 0 on (0,1). In case (b), Rychlik (1997) used the transformation $Y = -\ln(1-x)$ and transformed the problem to an equivalent problem of projecting $\psi_n(x) = \varphi_n(1-e^{-x}), x \geq 0,$ onto the cone of quantiles composed with the exponential distribution function in the inner product norm on R^+ with weight e^{-x} (cf., Gajek and Rychlik (1996)). Function φ_n as well as ψ_n are first strictly decreasing and ultimately strictly convex, increasing to infinity. Any nondecreasing concave function, say w is either nowhere greater than $\varphi_n(\psi_n)$ or $\varphi_n(\psi_n)$ and w cross each other at two points only. In the former case, the linear function is separating the graphs of $\varphi_n(\psi_n)$ and w is the best for the projection than w. If not so, the linear function is running through the crossing points. The linear function after determining the parameters (slope and Y-axis intercept) and minimizing the L^2 - distance, is unique best linear approximation which has the same zero integral. Both bounds in Theorem 5.1.2 are attained iff $Q(x)$ and $Q(1-e^{-x})$ are linear, i.e. for the uniform and exponential distributions, respectively.

The next theorem presented bounds for the ordinary record increments in terms of central absolute moments of various orders.

Theorem 5.1.3.

For c_p minimizing

$$\|P\varphi_n - c\|_q^q = \int_0^1 |P\varphi_n(x) - c|^q\, dx,$$

we have

$$EI_{n-1,n} \leq \left\| P\varphi_n - c_p \right\|_q \sigma_p .$$

The above bound is attained for

$$Q(x) = \mu + \frac{\sigma_p}{\left\| P\varphi_n - c_p \right\|_q^{q-1}} \left| P\varphi_n - c_p \right|^{q/p} \operatorname{sgn}\left(P\varphi_n - c_p\right), \quad \mu \in \mathrm{R}.$$

5.2. ORDINARY NONADJACENT RECORDS INCREMENTS

Recently Daniclak (2004) presented sharp upper bounds for expectation of increments of ordinary records (k = 1) measured in various scale units. Further she discussed the bounds under monotone density and monotone failure rate. Consider

$$EI_{m,n} = \int_0^1 \left(F^{-1}(x) - \mu\right)\left(\varphi_{m,n}(x) - c\right) dx,$$

for any arbitrary real c. We can observe

$$\varphi_{m,n}(x) = f_m(x) \left\{ \frac{m!\,[-\ln(1-x)]^{n-m}}{n!} - 1 \right\}.$$

Clearly $\varphi_{m,n}(x)$ is starting from 0 at 0, negative in (0, b), positive in (b, 1), where $b = 1 - e^{-(n!/m!)^{1/(n-m)}}$ and $\Phi_{m,n}(x)$ tends to $+\infty$ as $x \to 1$. One can also note that $\varphi_{m,n}(x)$ is decreasing in (0, a) and increasing in (a, 1), respectively, where $a = e^{-[(n-1)!/(m-1)!]^{1/(n-m)}}$. The antiderivate of $\varphi_{m,n}(x)$ can be written as

$$\Phi_{m,n}(x) = (x-1) \sum_{i=m+1}^{n} \frac{[-\ln(1-x)]^i}{i!}$$

Function $\Phi_{m,n}(x)$ is the concave decreasing in (0, a), convex decreasing in (a, b) and convex increasing in (b, 1). Also note that $\Phi_{m,n}(0) = \Phi_{m,n}(1) = 0$. Consider the line

$$l_\alpha(x) = \varphi_{m,n}(\alpha)(x-\alpha) + \Phi_{m,n}(\alpha),$$

tangent to $\Phi_{m,n}$ at $\alpha \in (a, b)$. By observing that the slopes of the above line continuously increase, value $l_\alpha(0)$ decreases, $l_\alpha(0) > \Phi_{m,n}(0)$ and $l_b(0) = \Phi_{m,n}(b) < 0$, we conclude that there exists a unique $\alpha^* \in (a,b)$ satisfying $l_{\alpha^*}(0) = 0$. Therefore the greatest minorant $\overline{\Phi}_{m,n}$ of $\Phi_{m,n}$ is given by

$$\overline{\Phi}_{m,n}(x) = \begin{cases} l_{\alpha^*}(n), & if\, 0 \le x < \alpha^* \\ \Phi_{m,n}(x), & if\, \alpha^* \le x < 1 \end{cases} \tag{5.2.1}$$

where α^* is the unique solution to

$$\alpha\, \varphi_{m,n}(\alpha) = \Phi_{m,n}(\alpha).$$

From (5.2.1), we obtain the projection of $\varphi_{m,n}$ as

$$P\varphi_{m,n}(x) = \varphi_{m,n}(\max\{\alpha^*, x\}).$$

Using the Moriguti's approach and the Hölder's inequality, we have

$$\begin{aligned} EI_{m,n} &\le \int_0^1 \left(F^{-1}(x) - \mu\right)\left(P\varphi_{m,n}(x) - \mu\right)\, dx \\ &\le \left\| P\varphi_{m,n} - c \right\|_q \sigma_p, \end{aligned} \tag{5.2.2}$$

for arbitrary constant $c = c^*$ minimizing the norm in (5.2.2). Considering $c = \varphi_{m,n}(d)$, $\alpha^* \le d < 1$, we can write the norm as follows

$$\begin{aligned} G(d) = \left\| P\varphi_{m,n} - \varphi_{m,n}(d) \right\|_q^q &= \int_0^1 \left| P\varphi_{m,n}(x) - \varphi_{m,n}(d) \right|^q dx \\ &= \int_0^{\alpha^*} \left(\varphi_{m,n}(d) - \varphi_{m,n}(\alpha^*)\right)^q dx + \int_{\alpha^*}^d \left(\varphi_{m,n}(d) - \varphi_{m,n}(x)\right)^q dx \\ &+ \int_d^1 \left(\varphi_{m,n}(x) - \varphi_{m,n}(d)\right)^q dx. \end{aligned}$$

Differentiation of $G(d)$ implies that there is a unique d^* that is minimizing the norm above. The latter inequality in (5.2.2) becomes equality if

$$\frac{F^{-1}(x) - \mu}{\sigma_p} = \frac{\left| P\varphi_{m,n}(x) - c^* \right|^{1/(p-1)} \operatorname{sgn}\left(P\varphi_{m,n}(x) - c^*\right)}{\left\| P\varphi_{m,n} - c^* \right\|_q^{1/(p-1)}}. \tag{5.2.3}$$

The RHS of (5.2.3) is non-decreasing and constant on all intervals on $\{x : \overline{\Phi}_{m,n}(x) \neq \Phi_{m,n}(x)\}$. This implies that the formula (5.2.3) satisfies the former equality in (5.2.2) as well. From (5.2.3), we have

$$\frac{F^{-1}(x)-\mu}{\sigma_p} = \begin{cases} -\left(\dfrac{c^* - P\varphi_{m,n}(\alpha^*)}{B_q}\right)^{1/(p-1)}, & \text{if } 0 \le x < \alpha^*, \\ -\left(\dfrac{c^* - P\varphi_{m,n}(x)}{B_q}\right)^{1/(p-1)}, & \text{if } \alpha^* \le x \le d, \\ \left(\dfrac{P\varphi_{m,n}(x) - c^*}{B_q}\right)^{1/(p-1)}, & \text{if } d \le x \le 1, \end{cases} \tag{5.2.4}$$

with $B_q = \| P\varphi_{m,n} - c^* \|_q$ From (5.2.4), we obtain the cdf $F(x)$ where the bound is attained. Note that the moment conditions are satisfied. The following theorem summarizes the result for $1 < p < \infty$.

Theorem 5.2.1.

If $X_1, X_2, \ldots,$ are i.i.d r.v.'s with continuous cdf F and $\sigma_p < \infty$, then

$$\frac{EI_{m,n}}{\sigma_p} \le B_q = B_q(m,n),\ 1 \le m < n,$$

where

$$B_q^q = \alpha^*(c^* - \varphi_{m,n}(\alpha^*))^q + \int_{\alpha^*}^{d}(c^* - \varphi_{m,n}(x))^q\,dx + \int_d^1(\varphi_{m,n}(x) - c^*)^q\,dx,$$

where $c^* = \varphi_{m,n}(d^*)$, *and* $d^* \in (\alpha^*, 1)$ *is the unique solution satisfying*

$$\alpha^*(\varphi_{m,n}(d) - \varphi_{m,n}(\alpha^*))^{q-1} + \int_{\alpha^*}^{d}(\varphi_{m,n}(d) - \varphi_{m,n}(x))^{q-1}\,dx$$
$$= \int_d^1(\varphi_{m,n}(x) - \varphi_{m,n}(d))^{q-1}\,dx.$$

The bound is attained for the distribution function of the form

$$F(x)=\begin{cases} 0, & if\ \dfrac{x-\mu}{\sigma_p}<-\left(\dfrac{c^*-\varphi_{m,n}(\alpha^*)}{B_q}\right)^{1/(p-1)}, \\ \varphi_{m,n}^{-1}\left(c^*-B_q\left(\dfrac{\mu-x}{\sigma_p}\right)^{p-1}\right), & if\ -\left(\dfrac{c^*-\varphi_{m,n}(\alpha^*)}{B_q}\right)^{1/(p-1)}\le\dfrac{x-\mu}{\sigma_p}<0, \\ \varphi_{m,n}^{-1}\left(c^*+B_q\left(\dfrac{\mu-x}{\sigma_p}\right)^{p-1}\right) & if\ \dfrac{x-\mu}{\sigma_p}\ge 0. \end{cases} \quad (5.2.5)$$

Distribution function in (5.2.5) has a jump at the left end of the support intervals. It is easily shown that we can find a sequence of continuous distributions $F_k, k\ge 1$ tending to the distribution function $F(x)$ in the desired norm. The continuity assumption is needed for the integral representation of the expectations of records. This means that the bound is attained in the limit by the distribution function given in (5.2.5).

If $p=2$, the constant c^* is found to be $c^*=0$. In this case, Raqab (2004a) established the following bound

$$\frac{EI_{m,n}}{\sigma}\le B_2,$$

where

$$B_2^2=\alpha^*\left[\frac{\left[-\ln(1-\alpha^*)\right]^n}{n!}-\frac{\left[-\ln(1-\alpha^*)\right]^m}{m!}\right]^2+(1-\alpha^*)\,[\binom{2n}{n}\sum_{i=0}^{2n}\frac{\left[-\ln(1-\alpha^*)\right]^i}{i!}$$

$$+\binom{n+m}{m}\sum_{i=0}^{m+n}\frac{\left[-\ln(1-\alpha^*)\right]^i}{i!}+\binom{2m}{m}\sum_{i=0}^{2m}\frac{\left[-\ln(1-\alpha^*)\right]^i}{i!}].$$

By Hölder's inequality, one can also obtain

$$EI_{m,n}\le\left\|\mathrm{P}\varphi_{m,n}-c\right\|_1\sigma_\infty. \quad (5.2.6)$$

The RHS of (5.2.6) can be simplified and written as

$$\left\|\mathrm{P}\varphi_{m,n}-c\right\|_1=(2d-1)\,\mathrm{P}\varphi_{m,n}(d)-2\overline{\Phi}_{m,n}(d).$$

Therefore $d^*=\max\{\alpha^*,1/2\}$ and the corresponding bound is

$$\frac{EI_{m,n}}{\sigma_\infty} \leq \sum_{i=m+1}^{n} \frac{(\ln 2)^i}{i!},$$

for $\alpha^* < 0.5$. If $a^* \geq 0.5$, the bound has to be

$$\frac{EI_{m,n}}{\sigma_\infty} \leq -\varphi_{m,n}(\alpha^*).$$

The bound in (5.2.6) is attained if $F^{-1}(x) - \mu$ takes positive and negative values on the sets where $\mathrm{P}\varphi_{m,n}(x) - \varphi_{m,n}(d^*)$ is negative and positive respectively. Also $F^{-1}(x) - \mu$ is constant on $[0, \alpha^*)$. From the moment conditions, the bound is attained in the limit for the marginal distribution

$$P(X_1 = \mu \pm \sigma_\infty) = 0.5$$

for $\alpha^* < 0.5$, and

$$P\left(X_1 = \mu - \frac{1-\alpha^*}{\alpha^*}\sigma_\infty\right) = \alpha^*, \quad P(X_1 = \mu + \sigma_\infty) = 1\text{-}\alpha^*,$$

for $\alpha^* \geq 0.5$.

The distribution function F is a decreasing (increasing) density distribution if $F^{-1}U$ is convex (concave) on (0,1), where U is the cdf of the standard uniform random variable. Distribution function F is decreasing (increasing) failure rate distribution if $F^{-1}V$ is convex (concave) on $(0, \infty)$, where V is the cdf of the standard exponential random variable. The expectation of the increment with $c = 0$ can be represented as

$$EI_{m,n} = \int_0^d \left(F^{-1}W(x) - \mu\right)\varphi_{m,n}W(x)\,w(x)\;dx, \tag{5.2.7}$$

where w(x) is the density function of $W = U, V$ on the support $(0, d)$. Arguments similar to those in Gajek and Rychlik (1996) and Rychlik (2001), Danielak (2004) presented the respective bounds. If the family of quantile functions $F^{-1}W(x) - \mu$ forms a convex cone, then the projections $P\varphi_{m,n}(x) = \varphi_{m,n}(\max\{\alpha^*, x\})$ and $P\varphi_{m,n}(V(x))$ are convex on their domains. This leads to the fact that the bounds in Theorem 5.2.1 are convex on their domains. This leads to the fact that the bounds in Theorem 5.2.1 are the desired bounds. The following Theorem presents the bounds for distributions with IFP and IFR.

Theorem 5.2.2.

If F has an IFP, then

$$\frac{EI_{m,n}}{\sigma_2} \leq \sqrt{3}\,\frac{2^{n-m}-1}{2^n}, \qquad (5.2.8)$$

and the bound is attained for the uniform distribution on $\left(\mu - \sqrt{3}\,\sigma_2,\ \mu + \sqrt{3}\,\sigma_2\right)$. *If the F has an IFR, then*

$$\frac{EI_{m,n}}{\sigma_2} \leq n - m, \qquad (5.2.9)$$

and the bound is attained for the exponential distribution with scale parameter σ_2 *and location parameter* $\mu - \sigma_2$.

Proof

The main idea is to find the projection of $\varphi_{m,n}W$, $W = U, V$ on the convex cone of non-decreasing concave functions, integrating to 0 on $(0,\ d)$ with respect to the weight function $w(x)$. As indicated in Rychlik (1997) that if function $\varphi_{m,n}$ is first strictly decreasing and ultimately strictly convex increasing, then the projection $p\varphi_{m,n}$ is linear function. It is easily checked that the linear approximations minimizing L^2 – distance to $\varphi_{m,n}W$ is

$$P\varphi_{m,n}(x) = 6\,\frac{2^{n-m}-1}{2^{n+1}}(2x-1),\ x \in (0,\ 1),$$

and

$$P\varphi_{m,n}V(x) = (n-m)\,(x-1),\ x \in (0,\ \infty).$$

The sharp bounds are just the norm of the above projections which are given in (5.2.8) and (5.2.9), respectively.

5.3. ADJACENT k^{th} INCREMENTS

In this section we present sharp moment bounds for the expectations of adjacent record increments $I_{n-1,n,k} = R_{n,k} - R_{n-1,k}$, $n = 1,2,\ldots$, and $k = 2, 3, \ldots$, based on combining the projection method (Moriguti, 1953) and Hölder inequality. The method is based on projection of some functions onto the family of nondecreasing functions. Raqab (2004a) established these bounds and determined the distribution functions where the bounds are achieved. Further, he presented the bounds numerically and compared them with other rough bounds for $p=1$, $1<p<\infty$, and $p=\infty$.

The expectation of the nth increment of kth RS's is

$$EI_{n-1,n,k} = \int_0^1 \left[F^{-1}(x)-\mu\right] \varphi_{n,k}(x)\ dx, \tag{5.3.1}$$

where $\varphi_{n,k}(x) = f^*_{n,k}(x) - f^*_{n-1,k}(x),\ 0 \le x \le 1$. Let us consider

$$EI_{n-1,n,k} = \int_0^\infty \left[F^{-1}\left(1-e^{-y}\right)-\mu\right] \left[h_{n,k}(y)-c\right] e^{-y}\, dy,\ n \ge 1,\ k > 1, \tag{5.3.2}$$

where $h_{n,k}(y) = g_{n,k}(y) - g_{n-1,k}(y), y > 0$, and c is an arbitrary constant. We omit case $n = k = 1$, because estimates for $EI_{0,1,1} = E(R_{1,1} - \mu)$ were already presented in Raqab and Rychlik (2002). The following three lemmas describe the variability of $\varphi_{n,k}(x)$ and its projection onto the cone of non-decreasing functions in $L^2([0,1], dx)$.

Lemma 5.3.1.

For $n = 2,3, \ldots$, and $k>1$, there exist $a_{n,k} \in \left[0,\ 1-e^{-n/k}\right)$ and $b_{n,k} \in \left(1-e^{-n/k}, 1\right)$ such that $\varphi_{n,k}(x)$ is decreasing on $\left[0,\ a_{n,k}\right)$ to $\varphi_{n,k}(a_{n,k}) < 0$, increasing on $\left(a_{n,k},\ b_{n,k}\right)$ to $\varphi_{n,k}(b_{n,k}) > 0$ and then decreasing on $\left(b_{n,k},\ 1\right)$ to 0.

Proof

Straightforward algebra leads to the following representation of $\varphi_{n,k}(x)$

$$\varphi_{n,k}(x) = -f^*_{n-1,k}(x) \left[\frac{k}{n}\ln(1-x)+1\right],\ n \ge 2.$$

For all $n \geq 2$, function $\varphi_{n,k}(x)$ starts from origin and vanishes as x approaches 1 passing the horizontal axis at $x = 1 - e^{-n/k}(n \geq 2, k > 1)$. By using the facts that

$$f_{n,k}^*(x) = (k/n)[-\ln(1-x)]\, f_{n-1,k}^*(x),$$

and

$$f_{n,k}^{*'}(x) = (k/n)\,[n + (k-1)\ln(1-x)]\,(1-x)^{-1} f_{n-1,k}^*(x),$$

we conclude that

$$\varphi_{n,k}'(x) = -\frac{k}{n-1} f_{n-2,k}^*(x)(1-x)^{-1}\left\{\frac{k(k-1)}{n}[-\ln(1-x)]^2 + (2k-1)\ln(1-x) + (n-1)\right\}. \tag{5.3.3}$$

It follows from (5.3.3) that $\varphi_{n,k}(x)$ decreases on $(0,\, a_{n,k})$, $(b_{n,k},\, 1)$ and increases on $(a_{n,k},\, b_{n,k})$, where $a_{n,k} = 1 - e^{-c_{n,k}}, b_{n,k} = 1 - e^{-d_{n,k}}$ with

$$c_{n,k} = \frac{(2k-1)n - \sqrt{(2k-1)^2 n + n(n-1)}}{2k(k-1)},$$

$$d_{n,k} = \frac{(2k-1)n + \sqrt{(2k-1)^2 n + n(n-1)}}{2k(k-1)}.$$

We can easily check that $\varphi_{n,k}(a_{n,k}) < 0$ and $\varphi_{n,k}(b_{n,k}) > 0$, which completes the proof.

Lemma 5.3.2.

For given n > 1 and k >1, there exists a unique $\alpha \in \lfloor c_{n,k}, n/k \rfloor$ *defined as the solution to equation*

$$[(k-1)y - n]e^{-y} = ky - n,\; n \geq 2,\; k > 1,$$

and $\beta = \frac{n}{k-1} \in \lfloor \frac{n}{k},\, d_{n,k} \rfloor$ *such that*

$$P\varphi_{n,k}(x) = \begin{cases} \varphi_{n,k}(1-e^{-\alpha}), & if\ x \le 1-e^{-\alpha}, \\ \varphi_{n,k}(x), & if\ 1-e^{-\alpha} < x < 1-e^{-\beta}, \\ \varphi_{n,k}(1-e^{-\beta}), & if\ x \ge 1-e^{-\beta}. \end{cases} \quad (5.3.4)$$

and every nondecreasing $w \in L^1([0,1], dx)$, *we have*

$$\int_0^1 w(x)\varphi_{n,k}(x)\, dx \le \int_0^1 w(x)\ P\varphi_{n,k}(x)\, dx \quad (5.3.5)$$

with the equality iff

$$w(x) = const,\ 0 < x < 1-e^{-\alpha},\ and\ 1-e^{-\beta} < x < 1. \quad (5.3.6)$$

Proof

From the statement of Lemma 5.3.1, we conclude that the antiderivative of $\varphi_{n,k}(x)$ needed for the Moriguti projection, $\Phi_{n,k}(x)$, is therefore concave decreasing, convex decreasing, convex increasing and concave increasing in $[0, a_{n,k}]$, $[a_{n,k}, 1-e^{-n/k}]$, $[1-e^{-n/k}, b_{n,k}]$, $[b_{n,k}, 1]$, respectively. Further, it is negative with $\Phi_{n,k}(0) = \Phi_{n,k}(1) = 0$. Thus its greatest convex minorant $\overline{\Phi}_{n,k}$ is linear in $[0, 1-e^{-\alpha}]$, and $[1-e^{-\beta}, 1]$ for some $\alpha \in [c_{n,k}, n/k]$ and $\beta \in [n/k, d_{n,k}]$. That is;

$$\overline{\Phi}_{n,k}(x) = \begin{cases} \varphi_{n,k}(1-e^{-\alpha})\, x + \Phi_{n,k}(0), & if\ x \le 1-e^{-\alpha}, \\ \Phi_{n,k}(x), & if\ 1-e^{-\alpha} < x < 1-e^{-\beta}, \\ -\varphi_{n,k}(1-e^{-\beta})(1-x) + \Phi_{n,k}(1), & if\ x \ge 1-e^{-\beta}, \end{cases}$$

where α and β are determined by solving the equations

$$H_{n,k}(y) = h_{n,k}(y)(1-e^{-y}), \quad (5.3.7)$$

and

$$H_{n,k}(y) = -h_{n,k}(y)\, e^{-y}, \quad (5.3.8)$$

where

$$H_{n,k}(y) = 1 - e^{-ky} \sum_{j=0}^{n} \frac{(ky)^j}{j!},$$

is the anti-derivative of $h_{n,k}(y)$ with respect to the exponential measure. Note that $\beta = n/(k-1)$ is the solution of Eq. (5.3.8). Eq. (5.3.7) has to be solved numerically in order to find the numbers α 's. The projection of $\varphi_{n,k}(x)$ onto the convex cone of non-decreasing functions in $L^2([0,1], dx)$ is given in (5.3.4). By the change of variables, we also deduce that the projection of $h_{n,k}(y)$ onto the analogous cone in $L^2([0,\infty), e^{-y}dy)$ has the following form:

$$Ph_{n,k}(y) = \begin{cases} h_{n,k}(\alpha), & \text{if } \; y \le \alpha, \\ h_{n,k}(y), & \text{if } \; \alpha < y < \beta, \\ h_{n,k}(\beta), & \text{if } \; y \ge \beta. \end{cases} \tag{5.3.9}$$

Lemma 5.3.3.

For given k>1, there exists $\beta = \frac{1}{k-1} \in \left[\frac{1}{k}, 1\right)$ *such that*

$$P\varphi_{1,k}(x) = \varphi_{1,k}(\min\{x, 1\text{-e}^{-1/(k-1)}\}), \tag{5.3.10}$$

and every non-decreasing $w \in L^1([0,1], dx)$, *we have*

$$\int_0^1 w(x)\, \varphi_{1,k}(x)\, dx \le \int_0^1 w(x) P\varphi_{1,k}(x)\, dx \tag{5.3.11}$$

with the equality iff

$$w(x) = const, \quad 1 - e^{-1/(k-1)} < x < 1. \tag{5.3.12}$$

Proof

Function $\varphi_{1,k}(x)$ is increasing-decreasing with the maximum at $1 - e^{-(2k-1)/k(k-1)}$ and passing the horizontal axis at $x = 1 - e^{-1/k}$. It starts from $-k$, and vanishes as x approaches 1. The anti-derivative is then convex decreasing, convex increasing and concave increasing in $\left[0, 1 - e^{-1/k}\right]$, $\left[1 - e^{-1/k}, 1 - e^{-(2k-1)/k(k-1)}\right]$, $\left[1 - e^{-(2k-1)/k(k-1)}, 1\right]$, respectively. It is also negative with $\Phi_{1,k}(0) = \Phi_{1,k}(1) = 0$. In fact, $\beta = 1/(k-1)$ is unique solution of (5.3.8) with n=1. Therefore (5.3.10) is the derivative of the greatest convex minorant of the anti-derivative of $\varphi_{1,k}(x)$. By Lemma 5.3.1, $P\varphi_{1,k}(x)$ is the projection of $\varphi_{1,k}(x)$ onto the family of non-decreasing functions in $L^2([0,1], dx)$. Also,

$$Ph_{1,k}(y) = h_{1,k}\left(\min\{y, 1/(k-1)\}\right), \tag{5.3.13}$$

is the analogous projection of $h_{1,k}(y)$ in $L^2\left([0,\infty), e^{-y}\, dy\right)$.

Putting $c = 0$ and applying directly the Hölder inequality to (5.3.2), we obtain

$$EI_{n-1,n,k} = \int_0^\infty \left[F^{-1}\left(1-e^{-y}\right) - \mu\right] h_{n,k}(y)\, e^{-y}\, dy$$

$$\leq \| F^{-1}(1-e^{-y}) - \mu \|_p \; \| h_{n,k} \|_q$$

$$= B_{n,k,p}(1)\, \sigma_p, \tag{5.3.14}$$

where $q = +\infty, p/(p-1)$, and 1 for $p = 1, 1 < p < +\infty$, and $p = +\infty$, respectively, and

$$\|h\|_r = \begin{cases} \left[\int_0^\infty |h(y)|^r e^{-y}\, dy\right]^{1/r}, & if \;\; 1 \leq r < +\infty, \\ ess \sup\limits_{y \in (0,\infty)} |h(y)|, & if \;\; r = +\infty, \end{cases}$$

denotes the norm of function h in the Banach space $L^r\left([0,\infty), e^{-y} dy\right)$. Accordingly, the bounds in (5.3.14) have the form

$$B_{n,k,p}(1) = \begin{cases} \sup\limits_{y \in (0,\infty)} |h_{n,k}(y)|, & if \;\; p = 1, \\ \left[\int_0^\infty |h_{n,k}(y)|^q e^{-y} dy\right]^{1/q}, & if \;\; 1 \leq p < +\infty, \\ \int_0^\infty |h_{n,k}(y)|\, e^{-y}\, dy, & if \;\; p + \infty. \end{cases} \tag{5.3.15}$$

In the special case $p = 2$, the bound in (5.3.15) reduces to non-sharp bound

$$B_{n,k,2}(1) = \left\{ k\left(\frac{k}{2k-1}\right)^{2n-1} \binom{2n-2}{n-1} + k\left(\frac{k}{2k-1}\right)^{2n+1} \binom{2n}{n} - 2k\left(\frac{k}{2k-1}\right)^{2n} \binom{2n-1}{n-1} \right\}^{1/2}.$$

Functions (5.3.9) and (5.3.13) are projections onto the convex cone of non-decreasing functions in $L^2([0,\infty), e^{-y}dy)$ (cf. Rychlik, 2001, pp. 14-16). Transformations of the projections are used for establishing improvements of bounds (5.3.15). Replacing $h_{n,k}$ by its projection $Ph_{n,k}$ and applying the Hölder inequality, we have

$$EI_{n-1,n,k} \le \int_0^\infty \left[F^{-1}(1-e^{-y})-\mu\right]\left[Ph_{n,k}(y)-c\right] e^{-y}dy$$

$$\le \| F^{-1}(1-e^{-y})-\mu \|_p \, \|P\, h_{n,k} \text{-c}\|_q$$

$$= \| Ph_{n,k} - c \|_q \, \sigma_{p,} \tag{5.3.16}$$

for arbitrary constant c. The equality in (5.3.16) holds if $F^{-1}(1-e^{-y})-\mu = w_p\left(Ph_{n,k}(y)-c\right)$, for some monotone function w_p and properly chosen constant c. The problem is that the monotone increase of $F^{-1}(1-e^{-y})-\mu$ is required, whereas $h_{n,k}(y)-c$ and accordingly $w_p(h_{n,k}(y)-c)$ are not monotone except the case $n=1,\ k=1$. In three subsequent theorems we analyze bounds for record increments measured in σ_p-units for $1<p<\infty, p=1$, and $p=+\infty$, respectively.

Theorem 5.3.1.

Let F be a continuous distribution function and $1<p<\infty$ *be fixed and* $q = p/(p-1)$. *For given* $n \ge 2$ *and* $k>1$, *we have*

$$EI_{n-1,n,k} \le B_{n,k,p}(2)\,\sigma_p, \tag{5.3.17}$$

where

$$B_{n,k,p}(2) = \| Ph_{n,k} - h_{n,k}(\eta^*) \|_{q,} \tag{5.3.18}$$

for unique η^* *satisfying equation*

$$\int_\alpha^\eta [h_{n,k}(\eta)-h_{n,k}(y)]^{q-1} e^{-y}\, dy + [h_{n,k}(\eta)-h_{n,k}(\alpha)]^{q-1}(1-e^{-\alpha}) =$$

$$\int_\eta^\beta [h_{n,k}(y)-h_{n,k}(\eta)]^{q-1} e^{-y}\, dy + [h_{n,k}(\beta)-h_{n,k}(\eta)]^{q-1} e^{-\beta}. \tag{5.3.19}$$

Equality in (5.3.17) holds for probability distribution

$$F(x)=\begin{cases}0, & if\ 0<x\le\xi_1,\\ \varphi_{n,k}^{-1}\left(h_{n,k}(\eta^*)+\operatorname{sgn}(x-\mu)B_{n,k,p}(2)\left|\frac{x-\mu}{\sigma_p}\right|^{p/q}\right), & if\ \xi_1<x<\xi_2,\\ 1, & if\ \xi_2\le x<1,\end{cases} \quad (5.3.20)$$

where

$$\xi_1 = \mu-\sigma_p\left[\frac{h_{n,k}(\eta^*)-h_{n,k}(\alpha)}{B_{n,k,p}(2)}\right]^{q/p},$$

$$\xi_2 = \mu+\sigma_p\left[\frac{h_{n,k}(\beta)-h_{n,k}(\eta^*)}{B_{n,k,p}(2)}\right]^{q/p}.$$

Proof

It is enough to minimize the latter norm in (5.3.16) with respect to $c=h_{n,k}(\eta),\ \eta\in(\alpha,\beta)$.

We have

$$\|Ph_{n,k}-h_{n,k}(\eta)\|_q^q=[h_{n,k}(\eta)-h_{n,k}(\alpha)]^q\left(1-e^{-\alpha}\right)$$

$$+\int_\alpha^\eta [h_{n,k}(\eta)-h_{n,k}(y)]^q e^{-y}dy$$

$$+\int_\eta^\beta [h_{n,k}(y)-h_{n,k}(\eta)]^q e^{-y}dy$$

$$+[h_{n,k}(\beta)-h_{n,k}(\eta)]^q e^{-\beta}. \quad (5.3.21)$$

Differentiation of the RHS of (5.3.21) and equating the result to 0 yields (5.3.19). It is clear to note that the LHS of (5.3.19) is continuous non-decreasing function in η, starting from 0, while the RHS is continuous function in η, starting from a positive value and vanishing at 0. This shows that there is a unique solution η^* of Eq. (5.3.19) that minimizes (5.3.21). Consequently, the optimal bound on $EI_{n-1,n,k}/\sigma_p$ is given by (5.3.18) that can be evaluated from (5.3.21) with $\eta=\eta^*$ defined by (5.3.19).

The latter inequality of (5.3.16) becomes equality iff

$$\tau = F^{-1}(1-e^{-y}) - \mu \,|\, Ph_{n,k}(y) - h_{n,k}(\eta^*)|^{q/p} \operatorname{sgn}\left(Ph_{n,k}(y) - h_{n,k}(\eta^*)\right) \tag{5.3.22}$$

for some $\tau \geq 0$. Clearly, (5.3.22) is non-decreasing and constant on all intervals of $\overline{H}_{n,k} < H_{n,k}$, which are contained in ones where $Ph_{n,k}$ is constant (see (5.3.6)), which is necessary and sufficient for the equality in the former inequality of (5.3.16). Condition $\int_0^\infty \left|F^{-1}(1-e^{-y}) - \mu\right|^p e^{-y} dy = \sigma_p^p$ forces

$$\tau = \frac{\sigma_p}{\left\|Ph_{n,k} - h_{n,k}(\eta^*)\right\|_q^{q/p}}. \tag{5.3.23}$$

Summing up, (5.3.22) with constants defined in (5.3.21) and (5.3.23) determines the unique distribution function for which

$$\frac{F^{-1}(x) - \mu}{\sigma_p} \,|^{p/q} \operatorname{sgn}(F^{-1}(x) - \mu) = \frac{\sigma_p}{B_{n,k,p}(2)} \left(\varphi_{n,k}(x) - h_{n,k}(\eta^*)\right).$$

This determines that the distribution function of the location-scale family for which the bound (5.3.18) is attained have the form given in (5.3.20). Distribution function (5.3.20) has two atoms of measures $(1-e^{-\alpha})$ and $e^{-n/(k-1)}$, respectively, at the ends of support intervals.

The following two theorems present sharp bounds of moment increments for the extreme cases of $p = 1$ and $p = +\infty$.

Theorem 5.3.2.

Let a continuous distribution function F have mean μ and first central absolute moment $\sigma_1 < +\infty$. Then for every $n \geq 2$ and $k > 1$,

$$\frac{EI_{n-1,n,k}}{\sigma_1} \leq B_{n,k,1}(2) = \frac{h_{n,k}(\beta) - h_{n,k}(\alpha)}{2}, \tag{5.3.24}$$

where α is uniquely determined by (5.3.7) *and $\beta = n/(k-1)$ is obtained from* (5.3.8). *The bound* (5.3.24) *is attained by three point marginal distribution*

$$F(x)=\begin{cases}0, & if\ \ x<\mu-\sigma_1\frac{1}{2(1-e^{-\alpha})},\\ 1-e^{-\alpha}, & if\ \ \mu-\sigma_1\frac{1}{2(1-e^{-\alpha})}\le x<\mu,\\ 1-e^{-\beta}, & if\ \ \mu\le x<\mu+\sigma_1\frac{e^{\beta}}{2},\\ 1, & if\ \ x\ge\mu+\sigma_1\frac{e^{\beta}}{2}.\end{cases}$$

Proof

By (5.3.2) and Lemma 5.3.2,

$$EI_{n-1,n,k}\le\int_0^\infty\left[F^{-1}(1-e^{-y})-\mu\right]\left[Ph_{n,k}(y)-c\right]e^{-y}dy$$

$$\le\sup_{x\in(0,\infty)}\left|Ph_{n,k}-c\right|\int_0^\infty\left|F^{-1}\left(1-e^{-y}\right)-\mu\right|e^{-y}dy$$

$$=\left\|Ph_{n,k}-c\right\|_\infty\sigma_1. \tag{5.3.25}$$

Since $Ph_{n,k}$ is nondecreasing function, the minimal value of the supremum is provided by $c=c^*=\lfloor P\varphi_{n,k}(0)+P\varphi_{n,k}(1)\rfloor/2$ and then the optimal bound has the form

$$B_{n,k,1}(2)=\frac{h_{n,k}(\beta)-h_{n,k}(\alpha)}{2}.$$

The first inequality in (5.3.25) becomes equality iff $F^{-1}(1-e^{-y})-\mu$ is constant on $\{y:Ph_{n,k}(y)=h_{n,k}(\alpha)\}$ and $\{y:Ph_{n,k}(y)=h_{n,k}(\beta)\}$. That is; the former bound of (5.3.25) is attained iff $F^{-1}(1-e^{-y})-\mu=c_1$ for $0<y<\alpha$ and $F^{-1}(1-e^{-y})-\mu=c_2$ for $\beta<y<\infty$. The latter bound is attained iff $F^{-1}(1-e^{-y})-\mu=0$ on $\{y:h_{n,k}(\alpha)<Ph_{n,k}(y)<h_{n,k}(\beta)\}$. The values of the atoms are deduced from the moment conditions $E(X_1-\mu)=0$ and $E|X_1-\mu|=\sigma_1$. As a consequence of that, the bound $B_{n,k,1}(2)$ is attained by three-point distribution function supported at $\mu-\sigma_1/2(1-e^{-\alpha}),\mu,\ \mu+\sigma_1 e^{\beta}/2$ with respective probabilities $1-e^{-\alpha}$, $e^{-\alpha}-e^{-\beta}$ and $e^{-\beta}$.

Theorem 5.4.3.

For a continuous distribution F on a finite interval $[\mu - \sigma_\infty, \mu + \sigma_\infty]$, *and arbitrary* $n \geq 2$ *and* $k > 1$, *we have*

$$\frac{EI_{n-1,n,k}}{\sigma_\infty} \leq B_{n,k,\infty}(2), \tag{5.3.26}$$

where

$$B_{n,k,\infty}(2) = \begin{cases} -g_{n-1,k}(\alpha)\left(\frac{k}{n}\alpha - 1\right), & \text{if } \alpha > \ln 2, \\ \frac{(k \ln 2)^n}{2^{k-1}\, n!}, & \text{if } \alpha \leq \ln 2 \leq \beta, \\ g_{n-1,k}(\beta)\left(\frac{1}{k-1}\right), & \text{if } \ln 2 > \beta. \end{cases} \tag{5.3.27}$$

Further, the bound (5.3.27) *is attained for the two-point marginal distributions with the following forms:*

$(i)\, For\, \alpha > \ln 2,$

$$P\left(X_1 = \mu + \left(1 - e^{\alpha}\right)^{-1}\sigma_\infty\right) = 1 - e^{-\alpha},\ P(X_1 = \mu + \sigma_\infty) = e^{-\alpha}.$$

$(ii)\, For\, \alpha \leq \ln 2 \leq \beta,$

$$P(X_1 = \mu \pm \sigma_\infty) = \frac{1}{2}$$

$(iii)\, For\, \beta < \ln 2,$

$$P(X_1 = \mu - \sigma_\infty) = 1 - e^{-\beta}, P\left(X_1 = \mu + \left(e^{\beta} - 1\right)\sigma_\infty\right) = e^{-\beta}.$$

Proof

Now

$$EI_{n-1,n,k} \leq \int_0^\infty \left[F^{-1}(1 - e^{-y}) - \mu\right]\left[Ph_{n,k}(y) - c\right] e^{-y} dy$$

$$\leq \sup_{x \in (0,\infty)} \left|F^{-1}\left(1 - e^{-y}\right) - \mu\right| \int_0^\infty \left|Ph_{n,k}(y) - c\right| e^{-y}\, dy$$

$$= \left\| Ph_{n,k} - c \right\|_1 \sigma_\infty, \tag{5.3.28}$$

Setting $c = h_{n,k}(\eta)$, for $\eta \in [\alpha, \beta]$, the norm in (5.3.28) can be analyzed as

$$\left\| Ph_{n,k} - c \right\|_1 = \left[h_{n,k}(\eta) - h_{n,k}(\alpha) \right] \left(1 - e^{-\alpha}\right) + \int_\alpha^\eta \left[h_{n,k}(\eta) - h_{n,k}(y) \right] e^{-y}\, dy$$

$$+ \int_\eta^\beta \left[h_{n,k}(y) - h_{n,k}(\eta) \right] e^{-y}\, dy + \left[h_{n,k}(\beta) - h_{n,k}(\eta) \right] e^{-\beta}$$

The optimal constant $c = h_{n,k}(\eta)$ that minimizes the above norm is $c = c^* = h_{n,k}(\eta^*)$, for $\eta^* = \alpha, \beta$, and $ln2$ if $\alpha > ln2$, $\beta < ln2$ and $\alpha \le ln2 \le \beta$, respectively. Summing up, (5.3.7), (5.3.8) with the above norm, the optimal bound is simplified to the bound $B_{n,k,\infty}(2)$ in (5.3.27).

The later bound in (5.3.28) becomes equality iff

$$F^{-1}\left(1 - e^{-y}\right) - \mu = \operatorname{sgn}\left(Ph_{n,k}(y) - c\right) \sigma_\infty,$$

on the intervals where $Ph_{n,k} - c$ does not vanish. This holds for the former inequality as well. Therefore, if $\alpha \le \eta^* = ln2 \le \beta$, the bound in (5.3.26) is attained if $F^{-1}(x) - \mu = \operatorname{sgn}(x - 1/2)\sigma_\infty$. If $\alpha > ln2$, the bound (5.3.26) is attained if $F^{-1}(x) - \mu = +\sigma_\infty$ for $1 - e^{-\alpha} < x < 1$, and is a constant otherwise. The first moment condition implies that F defines the two-point distribution supported on $\mu + \sigma_\infty / \left(1 - e^{\alpha}\right)$, and $\mu + \sigma_\infty$ with probabilities $1 - e^{-\beta}$ and $e^{-\beta}$, respectively. In the same manner, we can obtain analogous result for $\beta < ln2$.

For the special case $n = 1$ and $k > 1$, the projection of $\varphi_{n,k}$ reduced to $P\varphi_{1,k}$ described in (5.3.10). By combining Lemma 5.3.3 with the Hölder inequality, we present the following corollaries. However, the results are reductions of Theorems 5.3.1-5.3.3 by setting $n = 1$ and $\alpha = 0$ and observing $h_{1,k}(0) = -k$.

Corollary 5.3.1

Let a continuous distribution function F have mean μ and p^{th} central absolute moment σ_p^p, $1 < p < +\infty$. Then for every k>1,

$$\frac{EI_{0,1,k}}{\sigma_p} = \frac{E(R_{1,k} - X_{1,k})}{\sigma_p} \le B_{1,k,p}(2), \tag{5.3.29}$$

where

$$B_{1,k,p}(2) = \left\| Ph_{1,k} - h_{1,k}(\nu^*) \right\|_{q,}$$

for unique $\nu^* \in (1/k, 1/(k-1))$ *satisfying*

$$\int_0^{\nu} \lfloor h_{n,k}(\nu) - h_{n,k}(y) \rfloor^{q-1} e^{-y}\, dy = \int_{\nu}^{1/(k-1)} \lfloor h_{n,k}(y) - h_{n,k}(\nu) \rfloor^{q-1} e^{-y}\, dy$$

$$+ \left[h_{n,k}(\frac{1}{k-1}) - h_{n,k}(\nu) \right]^{q-1} e^{-\frac{1}{k-1}}. \tag{5.3.30}$$

Equality in (5.3.29) *holds for probability distribution*

$$F(x) = \begin{cases} 0, & \text{if } 0 < x \le \rho_1, \\ \varphi_{1,k}^{-1}(h_{1,k}(\nu^*) + \operatorname{sgn}(x-\mu) B_{1,k,p}(2) \left| \frac{x-\mu}{\sigma_p} \right|^{p/q}) & \text{if } \rho_1 < x < \rho_2, \\ 1, & \text{if } \rho_2 \le x < 1, \end{cases} \tag{5.3.31}$$

where

$$\rho_1 = \mu - \sigma_p \left[\frac{h_{1,k}(\nu^*) + k}{B_{1,k,p}(2)} \right]^{q/p},$$

$$\rho_2 = \mu - \sigma_p \left[\frac{h_{1,k}(\frac{1}{k-1}) - h_{1,k}(\nu^*)}{B_{1,k,p}(2)} \right]^{q/p},$$

Corollary 5.3.2.

Let a continuous distribution have mean μ *and first central moment* $\sigma_1 < +\infty$. *Then for every* $k > 1$,

$$\frac{EI_{0,1,k}}{\sigma_1}=\frac{E\left(R_{1,k}-X_{1,k}\right)}{\sigma_1}\leq B_{1,k,1}(2)=\frac{h_{1,k}(\frac{1}{k-1})+k}{2}. \tag{5.3.32}$$

The bound is attained for the three-point distributions supported at $\mu-\sigma_1 j/2$, μ, and $\mu+\sigma_1 e^{1/(k-1)}/2$ with respective probabilities $1/j, 1-1/j-e^{-1/(k-1)}$ and $e^{-1/(k-1)}$, as $j\to\infty$.

Corollary 5.3.3

For a continuous distribution function F on a finite interval $\left[\mu-\sigma_\infty,\mu+\sigma_\infty\right]$, *and arbitrary* $k>1$, *we have*

$$\frac{EI_{0,1,k}}{\sigma_\infty}=\frac{E\left(R_{1,k}-X_{1,k}\right)}{\sigma_\infty}\leq B_{1,k,\infty}(2), \tag{5.3.33}$$

where

$$B_{1,k,\infty}(2)=\begin{cases}\ln 2, & \text{if } k=2,\\ \frac{k}{k-1}e^{-1}, & \text{if } k=3,4,...,\end{cases}$$

For $k\geq 3$, *the bound is attained by two-point distribution supported on* $\mu-\sigma_\infty$ *and* $\mu+(e^{1/(k-1)}-1)\sigma_\infty$ *with probabilities* $1-e^{-1/(k-1)}$ *and* $e^{-1/(k-1)}$, *respectively. For* $k=2$ *the bound is attained by the distribution function symmetrically distributed at* $\mu\pm\sigma_\infty$.

Remark 5.3.1

The continuity assumption is needed for the integral representation of the expectations of kth record increments. Family of quantile functions of continuous distributions is not closed and the bounds on the integral are attained on the border. The bounds obtained here are sharp, because they are attainable in the limit by the quantiles of continuous distributions tending to the optimal function situated on the border. The solution of projection problem is unique and the limit is also unique for $1<p<\infty$. In the extreme cases $p=1$ and $p=\infty$, the bounds on the integral are attained in the limit by two-point or three-point distributions. In this representation, each element of any sequence of discrete distribution could be approximated by a continuous one.

We evaluate the p-norm bounds of the expectations of the nth value of the kth record increments $EI_{n-1,n,k}$, $1\leq n\leq 10$, $k=2,3,4,5$, in terms of absolute central population moments σ_p^p of order $p = 1,2,3,5$ and ∞. We use Mathematica 4.2 as a tool to determine numbers α 's and the values of the optimal constant $c = c^*$ by solving (5.3.7) and (5.3.19), respectively. The constants c^* 's are determined by means of the Newton-Raphson method. Then using (5.3.18), we compute the sharp bounds $B_{n,k,p}(2)$ for $p = 2,3,5$. These bounds are compared with the classical ones $B_{n,k,p}(1)$ described in (5.3.15) that are obtained without projection approach. The results are presented in Table 5.1.

For $p = 1$ and $p = \infty$, we use (5.3.24) and (5.3.27), respectively. Indeed by (5.3.15), we have

$$B_{n,k,1}(1) = \sup_{y\in(0,\infty)} |\varphi_{n,k}(y)| = \max\{-\varphi_{n,k}(a_{n,k}), \varphi_{n,k}(b_{n,k})\},$$

where

$$B_{n,k,\infty}(1) = 2\left[G_{n-1,k}(\frac{n}{k}) - G_{n,k}(\frac{n}{k})\right] = \frac{2\, n^n e^{-n}}{n!}.$$

On the other hand,

$$B_{1,k,1}(1) = \max\{k, \frac{k^2}{k-1}e^{-(2k-1)/k}\} = k > B_{1,k,1}(2) = \frac{k}{k-1}e^{-1}, k > 1,$$

$$B_{1,k,\infty}(1) = 2e^{-1} > B_{1,k,\infty}(2), \qquad (see\ (5.3.33)).$$

Unlike the projection bounds $B_{n,k,\infty}(2)$, the rough ones $B_{n,k,\infty}(1)$ are free of the parameter *k*. In Table 5.1, the numerical results show that the projection *p*-norm method obtained by combining the Moriguti approach with the Hölder inequality results in significant accuracy in evaluating the bounds. In fact, the *p*-norm bounds decrease as *p* increases which has the following explanation. If we consider σ_p as a general scale parameter and increase *p*, we restrict ourselves to narrower classes of distributions for which p^{th} moments exist, and the bounds in the narrower classes become tighter.

5.4. Non-Adjacent K^TH Increments

Recently Danielak and Raqab (2004a) discussed the bounds for non-adjacent k^{th} records. They provided a general set up for different cases of $I_{m,n,k} = R_{n,k} - R_{m,k}, k \geq 1$ and $0 \leq m < n$. They presented the bounds numerically in terms of mean absolute deviation

σ_1, standard deviation σ_2 and maximum absolute deviation σ_∞. The following Lemma describes the variability of

$$\varphi_{m,n,k}(x) = f^*_{n,k}(x) - f^*_{m,k}(x),\ 0 \le x \le 1,\ k \ge 1,\ 0 \le m < n$$

Table 5.1. Bounds on the kth Record Increments in Various Scale Units

		$p=1$		$p=2$		$p=3$		$p=\infty$	
k	n	$B_{n,k,1}(1)$	$B_{n,k,1}(2)$	$B_{n,k,2}(1)$	$B_{n,k,2}(2)$	$B_{n,k,3}(1)$	$B_{n,k,3}(2)$	$B_{n,k,\infty}(1)$	$B_{n,k,\infty}(2)$
2	1	2.0000	1.3679	0.8607	0.8521	0.8014	0.7881	0.7358	0.6931
	2	1.2360	0.7823	0.6285	0.6110	0.5878	0.5688	0.5413	0.4805
	3	1.9857	1.0460	0.6401	0.6160	0.5445	0.4967	0.4481	0.2996
	4	3.3999	1.6775	0.7213	0.6938	0.5489	0.4829	0.3907	0.2292
	5	6.0308	2.9036	0.8534	0.8239	0.5785	0.5025	0.3509	0.1922
	6	10.9397	5.2247	1.0388	1.0078	0.6266	0.5445	0.3212	0.1694
	7	20.1536	9.6131	1.2876	1.2554	0.6916	0.6050	0.2980	0.1538
	8	37.5493	17.9382	1.6163	1.5828	0.7736	0.6830	0.2792	0.1422
	9	70.5637	33.7961	2.0477	2.0128	0.8740	0.7794	0.2635	0.1333
	10	133.5040	64.1193	2.6122	2.5757	0.9953	0.8961	0.2502	0.1260
3	1	3.0000	1.7759	0.9675	0.9427	0.8525	0.8022	0.7358	0.5518
	2	0.8560	0.6334	0.6024	0.5681	0.5760	0.5569	0.5413	0.5405
	3	0.9890	0.5680	0.5291	0.4887	0.4932	0.4620	0.4481	0.3798
	4	1.2453	0.6316	0.5184	0.4752	0.4600	0.4125	0.3907	0.2742
	5	1.6321	0.7762	0.5365	0.4915	0.4490	0.3887	0.3509	0.2200
	6	2.1942	1.0085	0.5737	0.5275	0.4508	0.3815	0.3212	0.1874
	7	3.0024	1.3558	0.6270	0.5795	0.4614	0.3857	0.2980	0.1658
	8	4.1613	1.8639	0.6958	0.6472	0.4789	0.3985	0.2792	0.1504
	9	5.8245	2.6020	0.7810	0.7311	0.5024	0.4181	0.2635	0.1389
	10	8.2153	3.6722	0.8844	0.8331	0.5314	0.4437	0.2502	0.1298
4	1	4.0000	2.24525	1.0799	1.0415	0.9080	0.8258	0.7358	0.4905
	2	0.7966	0.6596	0.6293	0.5803	0.5913	0.5564	0.5413	0.4812
	3	0.8152	0.4981	0.5182	0.4636	0.4884	0.4568	0.4481	0.4440
	4	0.9003	0.4710	0.4768	0.4200	0.4397	0.3996	0.3907	0.3246
	5	1.0377	0.4962	0.4640	0.4061	0.4143	0.3639	0.3509	0.2530
	6	1.2291	0.5564	0.4671	0.4085	0.4017	0.3422	0.3212	0.2103
	7	1.4836	0.6493	0.4810	0.4217	0.3970	0.3304	0.2980	0.1822
	8	1.8159	0.7784	0.5034	0.4433	0.3979	0.3258	0.2792	0.1625
	9	2.2463	0.9513	0.5334	0.4723	0.4031	0.3267	0.2635	0.1479
	10	2.8019	1.1792	0.5705	0.5083	0.4119	0.3319	0.2502	0.1367
5	1	5.0000	2.7299	1.1858	1.1360	0.9602	0.8529	0.7358	0.4598
	2	1.3042	0.7214	0.6667	0.6053	0.6122	0.5592	0.5413	0.4229
	3	0.8527	0.4950	0.5300	0.4633	0.4955	0.4557	0.4481	0.4336
	4	0.7959	0.4271	0.4710	0.4028	0.4372	0.3966	0.3907	0.3757
	5	0.8535	0.4119	0.4428	0.3742	0.4038	0.3573	0.3509	0.2883
	6	0.9417	0.4241	0.4308	0.3621	0.3836	0.3301	0.3212	0.2356
	7	1.0598	0.4558	0.4289	0.3600	0.3716	0.3116	0.2980	0.2010
	8	1.2101	0.5044	0.4341	0.3649	0.3652	0.2995	0.2792	0.1768
	9	1.3971	0.5703	0.4448	0.3752	0.3626	0.2924	0.2635	0.1590
	10	1.6273	0.6553	0.4603	0.3901	0.3632	0.2892	0.2502	0.1454

Lemma 5.4.1.

Let $k > 1$. For $0 = m < n$, function $\varphi_{m,n,k}$ is increasing–decreasing. For $0 < m < n$, function $\varphi_{m,n,k}$ is decreasing–increasing–decreasing. Further $\varphi_{m,n,k}$ is negative in $(0, \eta_{m,n,k})$ and positive in $(\eta_{m,n,k}, 1)$ where

$$\eta_{m,n,k} = 1 - e^{-(q_{m,n,k})^{1/(n-m)}}, \text{ with } q_{m,n,k} = n!/(k^{n-m} m!).$$

Proof

Function $\varphi_{m,n,k}(x)$ is first negative and then positive (- +, for short). Therefore $\varphi_{m,n,k}(x)$ has exact one zero at η. Using the fact that the derivatives $f_{n,k}^{*'}(x)$ can be written as

$$f_{n,k}^{*'}(x) = \frac{k f_{n-1,k}^{*}(x) - (k-1) f_{n,k}^{*}(x)}{1-x},$$

we have

$$\varphi_{m,n,k}^{'}(x) = \frac{-k f_{m-1,k}^{*}(x) + (k-1) f_{m,k}^{*}(x) + k f_{n-1,k}^{*}(x) - (k-1) f_{n,k}^{*}(x)}{1-x}. \quad (5.4.2)$$

Applying Lemma 4.2.1 to (5.4.2) we determine the variability of function $\varphi_{m,n,k}$. We adopt the convention that $f_{j,k}^{*}(0)=0$, for $j < 0$. If $0 = m < n$, then function $\varphi_{m,n,k}^{'}$ is + - and then $\varphi_{m,n,k}$ is increasing-decreasing. For $0 < m < n$, function $\varphi_{m,n,k}^{'}$ is either - + -, or – on (0, 1). The latter one is impossible since $\varphi_{m,n,k}(0) = 0$ and $\varphi_{m,n,k}(x) \to 0$ as $x \to 1$. The projection of $\varphi_{m,n,k}$ is just the derivative of the greatest convex minorant, $\overline{\Phi}_{m,n,k}$, of $\Phi_{m,n,k}$ (Moriguti 1953). Next we determine the projections $\varphi_{m,n,k}$ for different cases.

Lemma 5.4.2

For given $k > 1$ and $0 < m < n$, the projection $P\varphi_{m,n,k}$ of the function $\varphi_{m,n,k}$ is

$$P\varphi_{m,n,k}(x)=\begin{cases}\varphi_{m,n,k}(\alpha), & \text{if } 0\le x\le\alpha,\\ \varphi_{m,n,k}(x), & \text{if } \alpha<x<\beta,\\ \varphi_{m,n,k}(\beta), & \text{if } \beta\le x<1,\end{cases} \tag{5.4.3}$$

where $\alpha\in(0,\eta_{m,n,k})$ *and* $\beta\in(\eta_{m,n,k},1)$ *are the unique solutions, respectively, to equations*

$$\Phi_{m,n,k}(x)= x\,\varphi_{m,n,k}(x), \tag{5.4.4}$$

$$\Phi_{m,n,k}(x)= \varphi_{m,n,k}(x)(x-1). \tag{5.4.5}$$

If $0=m<n$, *then*

$$P\varphi_{m,n,k}(x)=\varphi_{m,n,k}(\min\{\beta,x\}).$$

Proof

The projection $P\varphi_{m,n,k}$ *of* $\varphi_{m,n,k}$ onto the convex cone of non-decreasing function in $L^2([0,1],\, dx)$ is derived by means of Moriguti's greatest convex minorant method. First we consider the case $0<m<n.$ By Lemma 5.4.1, the anti-derivative $\Phi_{m,n,k}(x)$ starts from $\Phi_{m,n,k}(0)=0$, is concave decreasing in $(0,\,a)$ for some $a\in(0,\,\eta_{m,n,k})$, convex increasing in $(\eta_{m,n,k},\,b)$ for some $b\in(\eta_{m,n,k},\,1)$, concave increasing in $(b,\,1)$. It has global minimum at $\eta_{m,n,k}$ and $\Phi_{m,n,k}(\eta_{m,n,k})<0$ for $x\in(0,\,1)$. In this case, the greatest convex minorant $\overline{\Phi}_{m,n,k}(x)$ is involving two linear segments on the left and right ends of the domain. It is easily checked that there exist $\alpha\in(0,\,\eta_{m,n,k})$ and $\beta\in(\eta_{m,n,k},\,1)$ satisfying uniquely equations (5.4.4) and (5.4.5). It follows that the projection is as claimed in (5.4.3). If $0=m<n$, the $\Phi_{0,n,k}(x)$ is increasing convex, increasing concave and decreasing concave on $[0,\,b]$, $(b,\,\eta_{0,n,k})$, $(\eta_{0,n,k},\,1)$, where b is some constant in $(0,\,\eta_{0,n,k})$. It follows that the projection of $\varphi_{0,n,k}(x)$ is as claimed.

Next we use the lemmas 5.4.1 and 5.4.2 to present sharp upper bounds in terms of different scale units generated by various central absolute moments of a single observation. Applying (3.1.1) and Hölder's inequality, we obtain

$$EI_{m,n,k}\le\int_0^1\left[F^{-1}(x)-\mu\right]\left[P\varphi_{m,n,k}(x)-c\right]dx$$

$$\leq \left\| P\varphi_{m,n,k} - c \right\|_q \sigma_{p,} \qquad (5.4.6)$$

for arbitrary real c.

5.4.1. Case $1 < p < \infty$

Theorem 5.4.1.

Let $X_1, X_2, \ldots,$ *be a sequence of i.i.d r.v.'s with continuous* $df\ F, EX_1 = \mu$ *, and*

$\sigma_p^p = E|X_1 - \mu|^p < \infty,$ then

$$\frac{EI_{m,n,k}}{\sigma_p} \leq B_q(m,n,k), \qquad (5.4.7)$$

where

$$B_q^q(m,n,k) = \left\| P\varphi_{m,n,k} - c^* \right\|_q^q$$

$$= \alpha\left[c^* - \varphi_{m,n,k}(\alpha)\right]^q + (1-\beta)\left[\varphi_{m,n,k}(\beta) - c^*\right]^q$$

$$+ \int_\alpha^{d^*} [c^* - \varphi_{m,n,k}(x)]^q \, dx + \int_{d^*}^{\beta} | \varphi_{m,n,k}(x) - c^* |^q \, dx,$$

$c^* = \varphi(d^*)$ *and* $d^* \in (\alpha, \beta)$ *is the unique solution to the equation*

$$\alpha\left[\varphi_{m,n,k}(d) - \varphi_{m,n,k}(\alpha)\right]^{q-1} + \int_\alpha^d \left[\varphi_{m,n,k}(d) - \varphi_{m,n,k}(x)\right]^{q-1} dx =$$

$$(1-\beta)\left[\varphi_{m,n,k}(\beta) - \varphi_{m,n,k}(d)\right]^{q-1} + \int_d^\beta \left[\varphi_{m,n,k}(x) - \varphi_{m,n,k}(d)\right]^{q-1} dx \qquad (5.4.8)$$

The bound is attained in the limit by the distribution function having the form

$$F(x) = \begin{cases} 0, & \text{if } \frac{x-\mu}{\sigma_p} < \gamma_1, \\ \varphi_{m,n,k}^{-1}\left(c^* + \operatorname{sgn}(x-\mu) B_q(m,n,k) \left|\frac{x-\mu}{\sigma_p}\right|^{p-1} \right), & \text{if } \gamma_1 \leq \frac{x-\mu}{\sigma_p} < \gamma_2, \\ 1, & \text{if } x \geq \gamma_2, \end{cases} \qquad (5.4.9)$$

where

$$\gamma_1 = -\left(\frac{c^* - \varphi_{m,n,k}(\alpha)}{B_q(m,n,k)}\right)^{1/(p-1)}, \quad \gamma_2 = \left(\frac{\varphi_{m,n,k}(\beta) - c^*}{B_q(m,n,k)}\right)^{1/(p-1)}.$$

Proof

The constant c has to be chosen properly such that the norm in (5.4.6) is minimized. It suffices to consider $c = \varphi_{m,n,k}(d)$, where $d \in (\alpha, \beta)$. For this, let us define the function

$$Q(d) = \left\| P\varphi_{m,n,k} - \varphi_{m,n,k}(d) \right\|_q^q.$$

Its derivative can be represented as $Q'(d) = q\varphi'_{m,n,k}(d)K(d)$, where $K(d)$ increases on (α, β), $K(\alpha) < 0$ and $K(\beta) > 0$. It follows that there exists a unique root d^* of $Q'(d) = 0$ that minimizes $Q(d)$, which is a solution to (5.4.8). The latter inequality in (5.4.6) becomes equality if

$$\frac{F^{-1}(x) - \mu}{\sigma_p} = \frac{\left| P\varphi_{m,n,k}(x) - c^* \right|^{1/(p-1)} \operatorname{sgn}(P\varphi_{m,n,k}(x) - c^*)}{\left\| P\varphi_{m,n,k} - c^* \right\|^{1/(p-1)}} \tag{5.4.10}$$

The RHS of (5.4.10) is nondecreasing and constant on all intervals where $\overline{\Phi} \neq \Phi$. This implies that the formula (5.4.10) satisfies the former equality in (5.4.6) as well. From (5.4.10), we have

$$\frac{F^{-1}(x) - \mu}{\sigma_p} = \begin{cases} -\left(\frac{c^* - \varphi_{m,n,k}(\alpha)}{B_q(m,n,k)}\right)^{1/(p-1)}, & if \ 0 \le x \le \alpha, \\ -\left(\frac{c^* - \varphi_{m,n,k}(x)}{B_q(m,n,k)}\right)^{1/(p-1)}, & if \ \alpha < x \le d, \\ \left(\frac{\varphi_{m,n,k}(x) - c^*}{B_q(m,n,k)}\right)^{1/(p-1)}, & if \ d < x \le \beta, \\ \left(\frac{\varphi_{m,n,k}(\beta) - c^*}{B_q(m,n,k)}\right)^{1/(p-1)}, & if \ \beta < x \le 1. \end{cases} \tag{5.4.11}$$

It is easily checked that (5.4.11) integrates to *0* and its pth norm is 1. As a consequence of that the bound in (5.4.7) is attained by the distribution function given in (5.4.9). The distribution function $F(x)$ in (5.4.9) has two atoms of measures α and $1-\beta$, respectively, at the ends of support interval. Realizing the fact that the distribution function $F(x)$ is continuous iff its respective quantile function is strictly increasing, the quantile function of

$F(x)$ can be approximated by a sequence of strictly increasing functions $F_k^{-1}(x)$ with decreasing slopes to zero. That is; $F_k^{-1}(x)$ tends to $F^{-1}(x)$ in $L^2([0,1], dx)$ and then the expectations of kth records converge to the expectations with $F(x)$ defined in (5.4.9). The bounds obtained here are sharp, because they are attainable in the limit by the quantiles of continuous distributions tending to the optimal function situated on the border. The solution of projection problem is unique and the limit is also unique for $1 < p < \infty$.

For $p = q = 2$, we derive the bound expressed in standard deviation units σ_2.

Corollary 5.4.1.

Let $X_1, X_2, \ldots,$ be a sequence of i.i.d. r.v.'s with continuous df $F(x)$, $EX_1 = \mu$, and $\sigma_2^2 = Var(X_1) < \infty$, then

$$\frac{EI_{m,n,k}}{\sigma_2} \le B_2(m,n,k),$$

where

$$B_2^2(m,n,k) = \alpha\left[f_{n,k}^*(\alpha) - f_{m,k}^*(\alpha)\right]^2 + (1-\beta)\left[f_{n,k}^*(\beta) - f_{m,k}^*(\beta)\right]^2$$

$$+ k\left(\frac{k}{2k-1}\right)^{2m+1}\binom{2m}{m}\left[F_{2m,2k-1}^*(\beta) - F_{2m,2k-1}^*(\alpha)\right]$$

$$+ k\left(\frac{k}{2k-1}\right)^{2n+1}\binom{2n}{n}\left[F_{2n,2k-1}^*(\beta) - F_{2n,2k-1}^*(\alpha)\right]$$

$$- 2k\left(\frac{k}{2k-1}\right)^{m+n+1}\binom{m+n}{m}\left[F_{m+n,2k-1}^*(\beta) - F_{m+n,2k-1}^*(\alpha)\right], \tag{5.4.12}$$

where α and β are defined in (5.4.4) *and* (5.4.5). *The bound is attained in the limit by the distribution function $F(x)$ having the form*

$$F(x) = \varphi_{m,n,k}^{-1}\left(\tfrac{x-\mu}{\sigma_2} B_2(m,n,k)\right), \quad \frac{\varphi_{m,n,k}(\alpha)}{B_2(m,n,k)} \le \frac{x-\mu}{\sigma_2} < \frac{\varphi_{m,n,k}(\beta)}{B_2(m,n,k)}. \tag{5.4.13}$$

Proof

Using (5.4.4), (5.4.5) and (5.4.8), we find that $c^* = 0$. So the respective bound reduces to

$$B_2^2(m,n,k) = \alpha\varphi_{m,n,k}^2(\alpha) + (1-\beta)\ \varphi_{m,n,k}^2(\beta)$$

$$+ \int_\alpha^\beta \left[f_{m,k}^{*2}(x) + f_{n,k}^{*2}(x)(x) - 2 f_{m,k}^{*}(x) f_{n,k}^{*}(x) \right] dx.$$

Using the facts that

$$\int_a^b f_{i,k}^{*2}(x)\,dx = k(\frac{k}{2k-1})^{2i+1} \binom{2i}{i} \left[F_{2i,2k-1}^{*}(b) - F_{2i,2k-1}^{*}(a)\right],$$

and

$$f_{i,k}^{*}(x) f_{j,k}^{*}(x)\ dx = k(\frac{k}{2k-1})^{i+j+1} \binom{i+j}{i} f_{i+j,2k-1}^{*}(x),$$

we obtain (5.4.12). The bound is attained iff

$$F^{-1}(x) - \mu = \begin{cases} \frac{\sigma_2}{B_2(m,n,k)} \varphi_{m,n,k}(\alpha), \ if\ 0 \le x < \alpha, \\ \frac{\sigma_2}{B_{2(m,n,k)}} \varphi_{m,n,k}(x), \ if\ \alpha \le x < \beta, \\ \frac{\sigma_2}{B_2(m,n,k)} \varphi_{m,n,k}(\beta), \ if\ \beta \le x < 1, \end{cases}$$

from which we immediately obtain (5.4.13).

Now we consider the extreme cases $p = 1$ and $p = \infty$. The respective bounds are attained in the limit by two-point or three-point distributions. In this representation, each element of any sequence of discrete distribution could be approximated by a continuous one.

5.4.1. Case $p = 1$

Theorem 5.4.2.

Let $X_1, X_2, \ldots,$ be a sequence of i.i.d. r.v.'s, with continuous cdf, $EX_1 = \mu$ and $\sigma_1 = E|X_1 - \mu| < \infty$. With α and β defined in (5.4.4) and (5.4.5), we have

$$B_\infty(m,n,k) \le \frac{1}{2} \left\lfloor \varphi_{m,n,k}(\beta) - \varphi_{m,n,k}(\alpha) \right\rfloor. \tag{5.4.14}$$

The bound in attained in the limit by three-point marginal distribution

$$P\left(X=\mu-\frac{\sigma_1}{2\alpha}\right)=\alpha,\ P(X=\mu)=\beta-\alpha, P\left(X=\mu+\frac{\sigma_1}{2(1-\beta)}\right)=1-\beta.$$

Proof

From the monotonicity of $\varphi_{m,n,k}(x)$, the supremum in (5.4.6) is minimized by choosing $c^*=\left[P\,\varphi_{m,n,k}(0)+P\,\varphi_{m,n,k}(1)\right]/2$. So the respective bound has the form (5.4.14).

The first inequality in (5.4.6) becomes equality iff $F^{-1}(x)-\mu$ is constant on $(0,\alpha)$ and $(\beta,1)$. The latter becomes equality iff $F^{-1}(x)-\mu=0$ on the set $\{x:\alpha<x<\beta\}$. Therefore

$$\frac{F^{-1}(x)-\mu}{\sigma_1}=\begin{cases}c_{1,} & \text{if } 0\le x<\alpha,\\ 0, & \text{if } \alpha\le x<\beta,\\ c_{2,} & \text{if } \beta\le x<1,\end{cases}$$

where c_1 and c_2 are some negative and positive constants, respectively. From the moment conditions $E(X_1-\mu)=0$ and $E|X_1-\mu|=\sigma_{1,}$ we can easily conclude that the bound is attained by three-point distribution of support at $\mu-\sigma_1/2\alpha,\ \mu,\ \mu+\sigma_1/2(1-\beta)$ with respective probabilities $\alpha,\ \beta-\alpha,\ 1-\beta$, respectively.

5.4.2. CASE $p=\infty$.

Theorem 5.4.3.

Let X_1 bounded almost surely (a.s.) with continuous cdf F and α and β defined in (5.4.4) and (5.4.5), respectively. Then we have

$$B_1(m,n,k)=\begin{cases}-\varphi_{m,n,k}(\alpha), & \text{if } \alpha\ge\frac{1}{2},\\ \left(\frac{1}{2}\right)^{k-1}\sum_{j=m+1}^{n}\frac{(k\ln 2)^j}{j!}, & \text{if } \alpha<\frac{1}{2}<\beta,\\ \varphi_{m,n,k}(\beta), & \text{if } \beta\le\frac{1}{2},\end{cases} \tag{5.4.15}$$

The bound (5.4.15) is attained in the limit by the two-point marginal distribution defined by

$$P\left(X=\mu+\frac{\max\{1/2,\alpha\}-1}{\max\{1/2,\alpha\}}\sigma_\infty\right)=median\{1/2,\ \alpha,\ \beta\}$$

$$= 1 - P\left(X = \mu + \frac{\min\{1/2,\ \beta\}}{1 - \min\{1/2,\ \beta\}} \sigma_\infty \right). \tag{5.4.16}$$

Proof

Using (5.4.4), (5.4.5) and standard algebra, we obtain

$$\left\| P\,\varphi_{m,n,k} - \varphi_{m,n,k}(d) \right\|_1 = (2d-1)\,\varphi_{m,n,k}(d) - 2\Phi_{m,n,k}(d).$$

This leads to

$$d^* = \begin{cases} \alpha,\ if\ \alpha \ge \frac{1}{2}, \\ \beta,\ if\ \beta \le \frac{1}{2}, \\ \frac{1}{2},\ if\ \alpha < \frac{1}{2} < \beta. \end{cases}$$

If $\alpha < 1/2 < \beta$, the respective bound is $B_1(m,n,k) = -2\Phi_{m,n,k}(1/2)$. Further, $B_1(m,n,k) = -\varphi_{m,n,k}(\alpha)$, for $\alpha > 1/2$ and $B_1(m,n,k) = \varphi_{m,n,k}(\beta)$, for $\beta < 1/2$. The latter inequality in (5.4.6) becomes equality iff

$$F^{-1}(x) - \mu = \operatorname{sgn}(P\varphi_{m,n,k}(x) - c)\sigma_\infty,$$

on the intervals where $P\varphi_{m,n,k}(x) - c$ does not vanish. The former equality of (5.4.6) holds as well. If $\alpha \ge 1/2$, the bound in attained if $F^{-1}(x) - \mu = +\sigma_\infty$ for $\alpha < x < 1$, and a constant otherwise. The first moment condition implies that $F(x)$ defines the two-point marginal distribution as follows

$$F^{-1}(x) - \mu = \begin{cases} \frac{\alpha-1}{\alpha}\sigma_\infty,\ if\ 0 \le x < \alpha, \\ \sigma_\infty, \quad if\ \alpha \le x < 1. \end{cases}$$

Similarly, we can obtain analogous result for $\beta \le 1/2$. For $\alpha < 1/2 < \beta$, the bound is attained if $F^{-1}(x) - \mu = \operatorname{sgn}(x - 1/2)\sigma_\infty$. Summing up, the unified general form of the probability distribution is described in (5.4.15). By this, we complete the proof of Theorem 5.4.3.

In Table 5.2, we present the optimal sharp upper bound for $EI_{n,m,k}$, $k = 2, 3, 4, 5$ and different values of *m*, and *n*. The so obtained bounds are derived in terms of mean absolute deviation σ_1, standard deviation σ_2 and maximal absolute deviation σ_∞. For $k > 1,\ 0 = m < n$ and $k = 1,\ 0 < m < n,$ the results of Theorems 5.4.1-5.4.3 hold by setting

$\alpha = 0$ and $\beta = 1$, respectively. The bounds reduces to ones established by Danielak (2004). In the special case $k > 1$, $m = n-1$, the solution of (5.4.5) reduces to $\beta = 1 - e^{-n/(k-1)}$ and the respective bounds reduce to the bounds developed in Raqab (2004a). Case $k = 1, m = n-1, p = 2$ leads to simpler results by setting $\beta = 1$ (see, e.g., Rychlik, 2001, p. 140-141). If $n = 1$, $k > 1$, the increment becomes the difference between the 2nd value of kth records and the minimum of a sample of size $k, I_{0,1,k} = R_{1,k} - X_{1,k}$. The corresponding results are reductions of Theorem 5.4.1-5.4.3 by setting $\alpha = 0$, $\beta = 1/(k-1)$, with $\varphi_{0,1,k}(0) = -k$.

Table 5.2. Bounds on the kth Records Increments in Various Scale Units

			p=1	p=2	∞
k	m	n	$B_\infty(m,n,k)$	$B_2(m,n,k)$	$B_1(m,n,k)$
2	1	2	0.7823	0.6110	0.4805
	2	4	2.6326	1.2657	0.5047
	3	6	9.2963	2.3881	0.5473
	4	8	33.4815	4.3459	0.5935
	5	10	122.2410	7.7563	0.6379
3	1	2	0.6334	0.5681	0.5405
	2	4	1.1543	0.9279	0.6234
	3	6	2.2666	1.3998	0.6294
	4	8	4.6027	2.0575	0.6502
	5	10	9.5220	2.9840	0.6766
4	1	2	0.6596	0.5803	0.4812
	2	4	0.9304	0.8496	0.7518
	3	6	1.4230	1.1529	0.7267
	4	8	2.2635	1.5295	0.7245
	5	10	3.6828	2.0075	0.7335
5	1	2	0.7214	0.6053	0.4229
	2	4	0.8840	0.8324	0.8093
	3	6	1.1768	1.0621	0.8307
	4	8	1.6326	1.3269	0.8078
	5	10	2.3209	1.6420	0.8007

The continuity assumption actually is necessary for the integral representation of the expectation. The results indicate that the family of quantile functions of continuous distributions is not closed and the bounds on the integral are attained on the border. The bounds are sharp since they are attainable in the limit by the quantiles of continuous distributions tending to the optimal function situated on the border. The limit (i.e. solution of the projection problem) is unique and the sequences themselves are certainly not.

Danielak and Raqab (2004b) derived sharp upper bounds on expectations of increments of consecutive *k*th record statistics based on independent samples from restricted families of distributions: the ones with DFP and DFR. The results are obtained by means of projection method. Let $I_{n,k} = R_{n+1,k} - R_{n,k}$, $n \geq 0$, $k \geq 2$. In the following theorems we present sharp upper bound for $EI_{n,k}, n \geq 1, k \geq 2$ under the assumption that the parent distribution F is DFP. Let

$$\psi_{n,k}(x) = f^{*}_{n+1,k}(x) - f^{*}_{n,k}(x) \text{ and } \Psi_{n,k}(x) = F^{*}_{n+1,k}(x) - F^{*}_{n,k}(x).$$

Theorem 5.4.4.

Assume that $X_1, X_2, \ldots,$ are iid r.v.'s with distribution $F \in DFP$. Let α be a real constant satisfying

$$\alpha\, \psi_{n,k}(\alpha) = \Psi_{n,k}(\alpha).$$

Let υ be the smallest positive zero of

$$Y_1(x) = \sum_{i=0}^{n-2} \frac{3k^{n-i}}{(k+1)^{n-i+2}} f^{*}_{i,k}(x) + \left(k + \frac{3k}{(k+1)^3}\right) f^{*}_{n-1,k}(x)$$

$$+ \left(\frac{3}{(k+1)^2} - 2k + \frac{5}{2}\right) f^{*}_{n,k}(x) + \left(\frac{3}{k(k+1)} + k - \frac{5}{2}\right) f^{*}_{n+1,k}(x),$$

β be the smallest positive zero of

$$Z_1(x) = -\sum_{i=0}^{n-1} \frac{3k^{n-i}}{(k+1)^{n-i+2}} f^{*}_{i,k}(x) +$$

$$\left(\frac{1}{2} - \frac{3}{(k+1)^2}\right) f^{*}_{n,k}(x) + \left(\frac{2k-1}{k(k+1)} - \frac{1}{2}\right) f^{*}_{n+1,k}(x),$$

and

$$\lambda = \lambda_1(\beta) = \frac{3}{1-\beta}\left\{\frac{2-k}{2k} f^{*}_{n+1,k}(\beta) - f^{*}_{n,k}(\beta)\right\}$$

$$+ \frac{3}{(1-\beta)^3}\left(\frac{k}{k+1}\right)^{n+1}\left\{\frac{1}{k+1} - F^{*}_{n,k+1}(\beta) + \frac{k}{k+1} F^{*}_{n+1,k+1}(\beta)\right\}.$$

If

$$Y_1(\alpha) > 0 \ \textit{and} \ Z_1(\alpha) < 0 < Z_1(\upsilon),$$

then

$$\frac{EI_{n,k}}{\sigma} \leq B = B(n,k), \tag{5.4.17}$$

where

$$B^2(n,k) = \alpha\left[f^*_{n+1,k}(\alpha) - f^*_{n,k}(\alpha)\right]^2$$

$$+ k\left(\frac{k}{2k-1}\right)^{2n+3} \binom{2n+2}{n+1}\left[F^*_{2n+2,2k-1}(\beta) - F^*_{2n+2,2k-1}(\alpha)\right]$$

$$- 2k\left(\frac{k}{2k-1}\right)^{2n+2} \binom{2n+1}{n}\left[F^*_{2n+1,2k-1}(\beta) - F^*_{2n+1,2k-1}(\alpha)\right]$$

$$+ k\left(\frac{k}{2k-1}\right)^{2n+1} \binom{2n}{n}\left[F^*_{2n,2k-1}(\beta) - F^*_{2n,2k-1}(\alpha)\right]$$

$$+ (1-\beta)\left\{\left[f^*_{n+1,k}(\beta) - f^*_{n,k}(\beta)\right]^2 + \frac{1}{3}\lambda^2(1-\beta)^2\right.$$

$$\left. + \lambda(1-\beta)\left[f^*_{n+1,k}(\beta) - f^*_{n,k}(\beta)\right]\right\}. \tag{5.4.18}$$

The bound (5.4.17) is attained by

$$F(x) = \begin{cases} 0, & \frac{x-\mu}{\sigma} < \frac{\psi_{n,k}(\alpha)}{B}, \\ \psi^{-1}_{n,k}\left(\frac{x-\mu}{\sigma}B\right), & \frac{\psi_{n,k}(\alpha)}{B} \leq \frac{x-\mu}{\sigma} < \frac{\psi_{n,k}(\beta)}{B}, \\ \frac{x-\mu}{\sigma\lambda}B - \frac{\psi_{n,k}(\beta)}{\lambda} + \beta, & \frac{\psi_{n,k}(\beta)}{B} \leq \frac{x-\mu}{\sigma} < \frac{\lambda(1-\beta)+\psi_{n,k}(\beta)}{B}, \\ 1, & \frac{x-\mu}{\sigma} \geq \frac{\lambda(1-\beta)+\psi_{n,k}(\beta)}{B}. \end{cases} \tag{5.4.19}$$

Otherwise

$$\frac{EI_{n,k}}{\sigma} \leq \frac{f^*_{n+1,k}(\beta)}{k\,\beta}\sqrt{\frac{1+2\beta-3\beta^2}{3}}, \tag{5.4.20}$$

for the greatest $\beta \in (0,\, \alpha)$ satisfying

$$[k+1+3\beta(k-1)]f^*_{n+1,k+1}(\beta) = 6\beta\sum_{i=0}^{n} f^*_{i,k+1}(\beta).$$

The bound (5.4.20) is attained for

$$F(x)=\begin{cases} 0, & \dfrac{x-\mu}{\sigma}<u_1 \\ \dfrac{x-\mu}{2\sigma}(1-\beta)\sqrt{\dfrac{1+2\beta-3\beta^2}{2}}+\dfrac{1+\beta^2}{2}, & u_1\leq\dfrac{x-\mu}{\sigma}<u_2 \\ 1, & \dfrac{x-\mu}{\sigma}\geq u_2 \end{cases} \tag{5.4.21}$$

where $u_1 = (1-\beta)\sqrt{\dfrac{3}{1+2\beta-3\beta^2}}$, $u_2 = (1+\beta)\sqrt{\dfrac{3}{1+2\beta-3\beta^2}}$.

Distribution functions (5.4.21) are not absolutely continuous. However, one can find sequences of absolutely continuous cdf's $F_j \succeq_c U, j \to \infty$, ($F_j$ succeeds U in convex order) which achieve the bounds asymptotically.

Theorem 5.4.5.

Assume that $X_1, X_2, \ldots,$ are i.i.d r.v.'s with $F \in DFR$. Let V stand for standard exponential cdf and $\alpha_0 \in (0,\, V^{-1}(\theta))$ satisfies

$$\alpha_0\,\psi_{n,k}(\alpha_0) = \Psi_{n,k}(\alpha_0),$$

and v_0 be the smallest positive zero of

$$Y_1(x) = \frac{1}{2k^2}\sum_{i=0}^{n-2} f^*_{i,k}(x) + \frac{2k+1}{2k^2} f^*_{n-1,k}(x)$$

$$+\left(\frac{1}{2k^2}-2k+\frac{3}{2}\right)f_{n,k}^*(x)+\left(\frac{1}{2k^2}+k-\frac{3}{2}\right)f_{n+1,k}^*(x)$$

β_0 be the smallest positive zero of

$$Z_1(x)=-\sum_{i=0}^{n-1}f_{i,k}^*(x)+\left(k^2-1\right)f_{n,k}^*(x)-(k-1)^2 f_{n+1,k}^*(x),$$

and

$$\lambda=\lambda_1\left(V^{-1}(\beta_0)\right)=\frac{1}{2k^2}\sum_{i=0}^{n+1}f_{i,k}^*(\beta_0)+\frac{1}{2}f_{n,k}^*(\beta_0)-\frac{1}{2}f_{n+1,k}^*(\beta_0).$$

If

$$Y_1(\alpha_0)>0 \text{ and } Z_1(\alpha_0)<0<Z_1(v_0),$$

then

$$\frac{EI_{n,k}}{\sigma}\leq B=B(n,\,k), \tag{5.4.22}$$

where

$$B^2=\alpha_0\left[f_{n+1,k}^*(\alpha_0)-f_{n,k}^*(\alpha_0)\right]^2$$

$$+k\left(\frac{k}{2k-1}\right)^{2n+3}\binom{2n+2}{n+1}\left[F_{2n+2,2k-1}^*(\beta_0)-F_{2n+2,2k-1}^*(\alpha_0)\right]$$

$$-2k\left(\frac{k}{2k-1}\right)^{2n+2}\binom{2n+1}{n}\left[F_{2n+1,2k-1}^*(\beta_0)-F_{2n+1,2k-1}^*(\alpha_0)\right]$$

$$+k\left(\frac{k}{2k-1}\right)^{2n+1}\binom{2n}{n}\left[F_{2n,2k-1}^*(\beta_0)-F_{2n,2k-1}^*(\alpha_0)\right]$$

$$+(1-\beta_0)\{\left[f_{n+1,k}^*(\beta_0)-f_{n,k}^*(\beta_0)\right]^2+2\lambda^2$$

$$+2\lambda\left[f^*_{n+1,k}(\beta_0)-f^*_{n,k}(\beta_0)\right]\}.$$

The bound (5.4.22) is attained by

$$F(x)=\begin{cases} 0, & \text{if } \dfrac{x-\mu}{\sigma}<u_1^* \\ \psi^{-1}_{n,k}\left(\dfrac{x-\mu}{\sigma}B\right), & \text{if } u_1^*\le\dfrac{x-\mu}{\sigma}<u_2^* \\ V\left(\dfrac{x-\mu}{\sigma\lambda}B-\dfrac{\psi_{n,k}(\beta_0)}{\lambda}+V^{-1}(\beta_0)\right), & \text{if } \dfrac{x-\mu}{\sigma}\ge u_2^*, \end{cases}$$

(5.4.23)

with $u_1^*=\dfrac{\psi_{n,k}(\alpha_0)}{B}$, $u_2^*=\dfrac{\psi_{n,k}(\beta_0)}{B}$.

otherwise

$$\frac{EI_{n,k}}{\sigma}\le\frac{f^*_{n+1}V(\beta)}{k\left(e^{\beta}-1\right)}\sqrt{2e^{\beta}-1}, \tag{5.4.24}$$

for the greatest $\beta\in\left(0,V^{-1}(\alpha_0)\right)$ satisfying

$$k\left[1+V(\beta)+\beta\ V(\beta)\right]f^*_{n+1,k}V(\beta)$$

$$=V(\beta)\left(\sum_{i=0}^{n+1}f^*_{i+1,k}V(\beta)+(n+2)f^*_{n+2,k}V(\beta)\right).$$

The equality holds in (5.4.24) for

$$F(x)=\begin{cases} 0, & \dfrac{x-\mu}{\sigma}<-\dfrac{e^{-\beta}}{\sqrt{2e^{\beta}-1}} \\ V\left(\dfrac{x-\mu}{\sigma}\sqrt{2e^{\beta}-1}+e^{-\beta}+\beta\right), & \dfrac{x-\mu}{\sigma}\ge-\dfrac{e^{-\beta}}{\sqrt{2e^{\beta}-1}}. \end{cases}$$

Theorem 5.4.6.

Assume that $X_1, X_2, \ldots,$ *are i.i.d r.v.'s. If* $F \in DFP$, *then*

$$\frac{EI_{0,k}}{\sigma} = \frac{E(R_{1,k} - R_{0,k})}{\sigma} \leq \sqrt{3}\,\frac{k}{(k+1)^2},$$

and the equality is achieved for the uniform distribution on $(\mu - \sigma\sqrt{2}, \mu + \sigma\sqrt{3})$. *If* $F \in DFR$, *then*

$$\frac{EI_{0,k}}{\sigma} = \frac{E(R_{1,k} - R_{0,k})}{\sigma} \leq \frac{1}{k},$$

and the equality is achieved for the exponential distribution with location parameter $\mu - \sigma$ *and scale parameter* σ, *respectively.*

Chapter 6

CHARACTERIZATIONS BASED ON CONDITIONAL EXPECTATION

6.1. SOME KNOWN DISTRIBUTIONS

We will denote by $\zeta(\lambda, \gamma)$, the shifted exponential distribution with the pdf as given in the following form:

$$f(x) = \lambda e^{-\lambda(x-\gamma)}, \qquad x > \gamma, \lambda > 0$$

= 0, otherwise. (6.1.1)

Similarly, we denote by *POW* (θ,μ,ν) the power function distribution with the pdf as of the following form:

$$f(x) = \frac{\theta(\nu - x)^{\theta-1}}{(\nu-\mu)^{\theta}}, \quad -\infty < \mu < x < \nu < \infty$$

= 0 , otherwise. (6.1.2)

For the Pareto distribution, *PAR* (θ,μ,δ), we take the following pdf:

$$f(x) = \frac{\theta(\mu+\delta)^{\theta}}{(x+\delta)^{\theta}}, \theta > 0, x > \mu, \mu + \delta > 0$$

= 0 , otherwise. (6.1.3)

Using the conditional pdf of R_{m+k} given R_{m}. it can be shown that for the above three distributions

$$E(R_{m+k} | R_m = z] = a z + b, \tag{6.1.4}$$

for some constants a and b. For exponential distribution, $\zeta(\lambda,0)$, $a = 1$, $b = \lambda / k$. If X has power function distribution, *POW* $(\theta,0,1)$ with $F(x)=1-(1-x)^{\alpha}$, $0<x<1$, $\alpha>0$, then

$$a = \left(\frac{\theta}{\theta+1}\right)^k, \; b = 1 - \left(\frac{\theta}{\theta+1}\right)^k.$$

Also for Pareto distribution, *PAR* $(\theta,0,1)$, we have

$$a = \left(\frac{\theta}{\theta-1}\right)^k, \; b = \left(\frac{\theta}{\theta-1}\right)^k - 1.$$

6.2. Characterizations

Nagaraja (1977) obtained a characterization of these distributions using the relation (6.1.4). He gave the following theorem.

Theorem 6.2.1.

Let F be a continuous cumulative distribution function. If, for some constants a and b,

$E(R_n|R_{n-1}=x) = a\,x+b$, then except for a change of location and scale,

$F(x) = 1-(-x)^{\theta}$, $x<0$, if $0<a<1$

$F(x)= 1-e^{-x}$, $x \geq 0$, if $a=1$

$F(x)= 1-x^{\theta}$, $x>1$ if $a>1$,

where $\theta = a/(1-a)$. Here $a>0$.

Using the Theorem 6.2.1, the following theorem can easily be established.

Theorem 6.2.2.

Let F be a continuous cumulative distribution function. If, for some constants a and b,

$E(R_{n-1}|R_n=x) = a\,x+b$, then except for a change of location and scale,

$F(x) = 1-(-x)^{\theta}$, $x<0$, if $0<a<1$

$F(x)= 1-e^{-x}$, $x \geq 0$, if $a=1$

$F(x) = 1 - x^{\theta}$, $x > 1$ if $a > 1$,

where $\theta = a/(1-a)$. Here $a > 0$.

From Theorem 6.2.1 and 6.2.2, the following Theorem follows.

Theorem 6.2.3.

Let F be a continuous distribution function having both $E(R_n)$ and $E(R_{n+1})$ as finite. If the regression of $E(R_n | R_{n+1})$ and $E(R_{n+1}|R_n)$ are both linear in the conditioning variables, then F is an exponential distribution.

Ahsanullah and Wesolowski (1998) extended the result Theorem 6.2.1 to non adjacent record values. They characterized the exponential, power function and Pareto distributions by the following relation.

$$E(R_{m+2} \mid R_m) = a R_m + b,$$

where the constants a and b are of the following form:

(a) $a = 1$ and $b = \dfrac{2}{\lambda}$ for the exponential distribution, $\zeta(\lambda, \gamma)$,

(b) $a = \left(\dfrac{\theta}{\theta+1}\right)^2$ and $b = \dfrac{2\theta}{(\theta+1)^2}\nu$ for the power function distribution, $POW(\theta, \mu, \nu)$

(c) $a = \left(\dfrac{\theta}{\theta-1}\right)^2$ and $b = \dfrac{2\theta}{(\theta-1)^2}\delta$ for the Pareto distribution, $PAR(\theta, \mu, \delta)$, $\theta > 1$.

To prove these results, we need the following Lemma which is also of independent interest as a new characterization of the exponential distribution.

Lemma 6.2.1.

Let $\{X_n, n \geq 1\}$ be i.i.d. r.v.'s with absolutely continuous (with respect to Lebesgue measure) cdf F such that $\gamma = \inf\{x|F(x) > 0\} = 0$ and $\eta = \sup\{x|\, F(x) < 1\}$. Assume that $E(\exp(R_{m-1})) < \infty$. Then $X_n \in \zeta(\lambda, \gamma)$ iff $E[\exp(a R_{m+1})|R_{m-1}] = a \exp(c R_{m-1})$ with $\lambda = \dfrac{c\sqrt{a}}{\sqrt{a}-1}$ and $a \in (0,1)$ if $c < 0$ and $a > 1$ if $c > 0$.

Proof.

It is easy to verify that if $X_n \in \zeta(\lambda,\gamma)$ then $E[\exp(a R_{m+1})|R_{m-1}] = a \exp(c R_{m-1})$ with $\lambda = \dfrac{c\sqrt{a}}{\sqrt{a}-1}$ and $a \in (0,1)$ if $c < 0$ and $a > 1$ if $c > 0$.

Since $E(R_{m+1}) \geq E(R_{m-1})$ and we have $E[\exp(a R_{m+1})|R_{m-1}] = a \exp(c R_{m-1})$, then it follows easily that $a \in (0,1)$ if $c < 0$ and $a > 1$ if $c > 0$. Using the relation

$E[\exp(a R_{m+1})|R_{m-1}] = a \exp(c R_{m-1})$, we get from (1.2.12)

$$\int_x^\infty e^{cy} \{R(y) - R(x)\} f(y)\, dy = a e^{cx} \overline{F}(x), \tag{6.2.1}$$

Differentiating both sides of (6.2.1) with respect to x and simplifying, we have

$$\int_x^\infty e^{cy} f(y)\, dy = ac\, e^{cx} \frac{(\overline{F}(x))^2}{f(x)} - a\, e^{cx} \overline{F}(x). \tag{6.2.2}$$

If we differentiate both sides of (6.2.2) with respect to x and simplify, then we obtain the following equation.

$$f'(x) \overline{F}^2(x) - cf(x)\overline{F}^2(x) + 3\overline{F}(x) f^2(x) + \frac{1-a}{ac} f^3(x) = 0. \tag{6.2.3}$$

Let $y = \overline{F}(x)$ (i.e. $f(x) = -y'$, $f' = -y''$), then (6.2.3) yields

$$-y'' y^2 + ay' y^2 + 3(y')^2 y - \frac{1-a}{ac}(y')^3 = 0. \tag{6.2.4}$$

Substituting $u(y) = y'$ in (6.2.4), we have

$$-u' y^2 + ay^2 + 3u\, y - \frac{1-a}{ac} u^2 = 0. \tag{6.2.5}$$

Let $u(y) = w(y) - \lambda y$, where λ is a real constant. Rewriting the equation (6.2.5) in terms of w, we get

$$-w' y^2 + \lambda y^2 + a\, y^2 + 3wy - 3\lambda y^2 - \frac{1-a}{ac}(w^2 - 2\lambda y + \lambda^2 y^2) = 0 \tag{6.2.6}$$

We choose λ in such that

$$\frac{1-a}{ac}\lambda^2+2\lambda \ -c=0. \tag{6.2.7}$$

Let λ_1 and λ_2 be the two possible roots of λ of the equation (6.2.7). Then

$\lambda_1=\dfrac{c\sqrt{a}}{\sqrt{a}-1}$ and $\lambda_2=\dfrac{c\sqrt{a}}{\sqrt{a}+1}$. With λ_1 or λ_2, we have from (6.2.6),

$$y^2\, w'+(3-2\lambda\delta)\, wy-\delta\, w^2=0, \tag{6.2.8}$$

where $\delta=\dfrac{1-c}{ac}$.

The equation (6.2.8) is of Bernoulli type. Take $\lambda=\lambda_1$. If $w \equiv 0$, then $y' = u(y) = -\lambda_1 y$ and hence

$$y=\overline{F}(x)=k\, e^{-\lambda_1 x} \tag{6.2.9}$$

The boundary conditions $\overline{F}(\gamma) = 1$ and $\lim_{x\to\nu}\overline{F}(x) = 0$, imply that $\nu = \infty$ and

$$y=\overline{F}(x)=\ e^{-\lambda_1(x-\gamma)}\ ,\ x\geq\gamma. \tag{6.2.10}$$

The solution of (6.2.10) for $w \not\equiv 0$, is

$$w=\frac{2(1+\lambda_1\delta)\ y^{2(1+\lambda_1\delta)+1}}{2D(1+\lambda_1\delta)+\delta y^{2(1+\lambda_1\delta)}}\ , \tag{6.2.11}$$

where D is an arbitrary constant. Writing in terms of y, we have

$$-y'=y\frac{\lambda_1 Dy^B+C}{Dy^B+A}\ , \tag{6.2.12}$$

where $A=\dfrac{\delta}{2(1+\lambda_1\delta)}$, $C = \lambda_1 A - 1$, $B = - d/A$. The equation (6.2.12) implies

$$f(x)=\overline{F}(x)\frac{\lambda_1 D(\overline{F}(x))^B+C}{D(\overline{F}(x))^B+A}\ . \tag{6.2.13}$$

Let us consider all possible solutions.

First case: c < o, then a ∈ (0,1) and λ_1 and λ_2 are of opposite sign ($\lambda_2 < 0 < \lambda_1$). If w ≡ 0, then from (6.2.13), we have $\overline{F}(x)=k\,e^{-\lambda_1 x}$. Since F is a probability distribution function, we must have

$$\overline{F}(x)=e^{-\lambda_1(x-\gamma)}, x\geq\gamma , \lambda_1 >0. \tag{6.2.14}$$

If $w \neq 0$, then A > 0, C <0 and B >0. Thus if x → η, then the right hand side of (6.2.13) is negative but the left hand side of (6.2.13) is always positive. Thus the only solution for c < 0 is the one given by (6.2.18).

Second case: c > 0, then a > 1 and both the roots are positive ($\lambda_1 > \lambda_2 > 0$). If w ≡ 0, then as in the first case the only solution is given by (6.2.14) . Observe that λ_1 > c, hence the integrability condition is satisfied.

.

If $w \neq 0$, then A > 0, B >0 and C > 0. Thus from (6.2.13), we get

$$\overline{F}(x) = e^{-\lambda_2 x}\left(\frac{\lambda_1 D\overline{F}^{B} + C}{\lambda D + C}\right)^{-\frac{1}{\gamma\delta}} \tag{6.2.15}$$

Let H(D,y) be the right hand side of (6.2.15). If D < 0, then H(D,y) ≥ H(D,1) and if D > 0, then H(D,y) ≥ H(D,0). Let M_0 = min { H(D,1) and H(D,0) }. Thus $\overline{F}(x)\geq M_0\, e^{-\lambda_2 x}$.

The condition of the finiteness of $E(e^{cX_n})$ implies that λ_2 > c. Since $\lambda_2 = \frac{c\sqrt{a}}{\sqrt{a}+1}$ < c, we have the contradiction. Hence the only solution of (6.2.13) is the one given by (6.2.14). We now give the characterization theorem.

Theorem 6.2.4.

Let $\{X_n, n \geq 1\}$ be i.i.d. r.v.'s with absolutely continuous (with respect to Lebesgue measure) cdf F such that γ = inf {x|F(x) > 0} = 0 and η = sup { x| F(x) < 1} < ∞. Assume that for some δ > 0, $E(X^{1+\delta})$ exists. If $E(R_{m+1} \mid R_{m-1}) = a R_{m-1} + b$, then only the following three cases are possible:

1. If a < 1, then $X_n \in POW$ (θ,μ,ν), where $\theta=\frac{\sqrt{a}}{2-\sqrt{a}}, \nu=\frac{b}{1-a}$ and μ is an arbitrary real number with μ< ν.
2. If a > 1, then $X_n \in PAR$ (θ,μ,δ), where $\theta=\frac{\sqrt{a}}{\sqrt{a}-1}, \delta=\frac{b}{a-1}$ and $\mu>\frac{b}{1-a}$.

3. If a = 1, then $X_n \in \zeta(\lambda,\gamma)$, where $\lambda = \frac{2}{b}$ and γ is an arbitrary real number.

Proof.

Suppose a < 1. Let $\gamma = \inf\{x|F(x) > 0\}$ and $\eta = \sup\{x|F(x) < 1\}$ Then the assumption that $E(R_{m+1} | R_{m-1}) = a R_{m-1} + b$ implies that for all x, $x \le ax + b$. Thus $\eta = \frac{b}{1-a} < \infty$. Let us define the random variable Z_n by $Z_n = -\ln(\eta - X_n)$, $n \ge 1$. Let $V_1, V_2, \ldots,$ be the record values of Z_n,
$n \ge 1$. Then the condition $E(R_{m+1} | R_{m-1}) = a R_{m-1} + b$ reduces to $E(e^{V_{m+1}} | V_{m-1}) = a e^{V_{m-1}}$. Hence by Lemma 6.2.1, $Z_n \in \zeta(\lambda,\eta)$, where $\lambda = \frac{\sqrt{a}}{1-\sqrt{a}}$ and γ is an arbitrary number. Thus

$X_n \in POW(\theta,\mu,\gamma)$, where $\theta = \frac{\sqrt{a}}{\sqrt{a}-1}$, γ is as given above and $\mu = \gamma - e^{-\eta}$.

Consider a > 1. Let $\gamma = \inf\{x|F(x) > 0\}$ and $\eta = \sup\{x|F(x) < 1\}$. Let $\delta = \frac{b}{a-1}$ and consider the record values of $Z_n = \ln(\delta + X_n)$. Then the condition $E(R_{m+1} | R_{m-1}) = a R_{m-1} + b$ reduces to $E(e^{V_{m+1}} | V_{m-1}) = a e^{V_{m-1}}$. Hence by Lemma 2.1, $Z_n \in \zeta(\lambda,\eta)$, where $\lambda = \frac{\sqrt{a}}{1-\sqrt{a}}$ and γ is an arbitrary number. Thus $X_n \in PAR(\theta,\mu,\gamma)$ where

$\theta = \frac{\sqrt{a}}{\sqrt{a}-1}$, $\delta = \frac{b}{a-1}$ and $\mu = e^{\gamma} - \delta$.

Consider a = 1, then using (6.2.12), we get

$$\int_x^{\eta} y\{R(y) - R(x)\} f(y)\,dy = (x+b)\bar{F}(x). \tag{6.2.16}$$

Differentiating both sides of (6.2.16), we get

$$\int_x^{\eta} y f(y)\,dy = -(x+b)\bar{F}(x) + \frac{(\bar{F}(x))^2}{f(x)}. \tag{6.2.17}$$

Since $\int_x^{\eta} y f(y)\,dy = -x\bar{F}(x) - \int_x^{\eta} \bar{F}(y)\,dy$, we have from (6.2.17)

$$\int_x^{\eta} \overline{F}(y)dy = -b\overline{F}(x) + \frac{(\overline{F}(x))^2}{f(x)} \tag{6.2.18}$$

Differentiating both sides of (6.2.18), we respect to x, we obtain on simplification

$$\overline{F}^2(x)\, f'(x) + 3\overline{F}(x)\, f^2(x) - b\, f^3(x) = 0. \tag{6.2.19}$$

Now substituting $y = \overline{F}(x)$ (i.e. $f(x) = -y'$, $f' = -y''$) the equation (6.2.19) reduces to

$$y''\, y^2 + 3y\, y'^2 + 2b\, y'^3 = 0. \tag{6.2.20}$$

Upon substitution $u(y) = y'$ the equation (6.2.20) reduces to the Bernoulli type finally yielding

$$y' = u = \frac{2y^3}{2D - by^2}, \tag{6.2.21}$$

where D is an arbitrary constant. Rewriting (6.2.21), we get

$$f(x) = \frac{2(\overline{F}(x))^3}{b(\overline{F}(x))^2 - D}. \tag{6.2.22}$$

Since $f(x) \geq 0$ for all x and $\overline{F}(x) \to 0$ as $x \to \eta$, it follows that $D \leq 0$. Finally the solution of (6.2.22) is given by

$$-\frac{b}{2}\ln \overline{F}(x) - \frac{D}{2}(\overline{F}(x))^{-2} = x + d \tag{6.2.23}$$

where d is an arbitrary real number. Since $\lim_{x\to\gamma} \overline{F}(x) = 1$, we have $\gamma > -\infty$ and $\gamma + d = -D/2 \geq 0$.

Thus we can write (6.2.23) as

$$\overline{F}(x) = \exp\left[-\frac{2}{b}(x + \mu\right] \tag{6.2.24}$$

Suppose $\eta < \infty$, then $\lim_{x\to\eta} \overline{F}(x) = 0$ but the right hand side of (2.2.24) remains positive, hence $\eta = \infty$. Now by Markov's inequality

$\overline{F}(x) \le \frac{x}{m}$ for sufficiently large positive x, where m = E(|X_n|).

Thus for large x we have

$$\overline{F}(x) \ge \exp\left[-\frac{2}{b}(x+d) + \frac{2}{b}(\gamma+d)\frac{x^2}{m^2} \right] \quad (6.2.25)$$

If γ+d > 0, then the right hand side of (6.2.25) is unbounded for x →∞ which contradicts the assumption that $0 \le \overline{F}(x) \le 1$. Hence γ+d = 0 and thus from (6.2.24), we get

$$\overline{F}(x) = \exp\left[-\frac{2}{b}(x+\gamma) \right]. \quad (6.2.26)$$

Lopez-Blazquez and Moreno-Rebollo (1997) gave characterizations of distributions by using the following linear property

$$E(R_{k+s} | R_k = z] = a\,z + b. \quad (6.2.27)$$

for1<k<s<n. Their result is given in the following theorem.

Theorem 6.2.5.

Let X be a random variable with distribution function F(x) such that 0<F(x)<1 for all x. We assume further that F(x) is k-times differentiable for all x and

$$E(R_{k+i} | R_i = z] = az + b. \quad (6.2.28)$$

Then except for location and scale parameters,

$$F(x) = 1 - |x|^{\delta}, \text{for } x \in [-1,0], \text{if } 0<a<1 \quad (6.2.29)$$

$$F(x) = 1\text{-}\exp(-x), \text{for } x \in [0, \infty), \text{if } a=1 \quad (6.2.30)$$

$$F(x) = 1\text{-}x^{\delta}, \text{ for } x \in [1, \infty), \text{if } a>1 \quad (6.2.31)$$

where $\delta = (r-(n-i))^{-1}$ and r is the unique real root greater than k-1 of the polynomial equation

$P_k(x) = P_k(n\text{-}i) / b$,

and

$P_k(x) = x(x-1)(x-2)\ldots(x-k+1)$.

Lopez-Blazquez and Moreno-Rebollo (1997) also gave characterizations of distributions by using the following linear property

$$E(R_k | R_{k+s} = z] = az + b, \ 1 \leq k<s<n.$$

Raqab (2002) and Wu (2004) considered this problem for non-adjacent record values under some stringent smoothness assumptions on the distribution function $F(.)$. Dembinska and Wesolowski (2000) characterized the distribution by means of the relation

$$E(R_{k+s} | R_k = z] = az + b, \text{ for } 1<k<s<n.$$

They used a result of Rao and Shanbhag (1994) which deals with the solution of extended version of integrated Cauchy functional equation. It can be pointed out earlier that Rao and Shanbhag's result is applicable only when the conditional expectation is a linear function

We shall attempt to characterize distributions by means of the relation.

$$E(\psi(R_{k+s}) | R_k = z) = g(z), k, s \geq 1,$$

where $k,s \geq 1$ and $\psi(x)$ is a continuous function.

The following theorem is due to Gupta and Ahsanullah (2004).

Theorem 6.2.6.

Let $X_j = 1,2,3\ldots,$ *be a sequence of independent identically distributed random variables with absolutely continuous (w.r.t. Lebesgue measure) distribution F(x) and probability density function f(x). Let* R_j, *j=1,2,3,…, be a sequence of upper record values, then the condition*

$$E[\psi(R_{k+s}) | R_k = z] = g(z), \tag{6.2.32}$$

where $k, s \geq 1$, *and* $\psi(x)$ *is a continuous function, determines the distribution uniquely.*

Proof

It can be verified that

$$E\left[\psi(R_{k+s})\middle|R_k = z\right] = \int_z^\infty \frac{\psi(x)\left[R(x) - R(z)\right]^{s-1}}{\overline{F}(z)} d(-\overline{F}(x)), \tag{6.2.33}$$

where $R(x) = -\ln \overline{F}(x)$.

Case: s = 1

In this case, using the above two equations, we obtain

$$\int_z^\infty \psi(x) f(x) dx = g(z)\overline{F}(z). \tag{6.2.34}$$

Differentiating both sides of (6.2.34) with respect to z and simplifying, we obtain

$$r(z) = \frac{f(z)}{\overline{F}(z)} = \frac{g'(z)}{g(z) - \psi(z)}, \tag{6.2.35}$$

where $r(z)$ is the failure rate of the original distribution. Hence the result follows.

Case: s = 2

In this case, we obtain

$$\int_z^\infty \psi(x)\left[R(x) - R(z)\right] f(x) dx = g(z)\overline{F}(z) \tag{6.2.36}$$

Differentiating both sides of (6.2.36) with respect to z, we obtain

$$-\int_z^\infty \psi(x) f(x) dx = g'(z)\frac{(\overline{F}(z))^2}{f(z)} - g(z)\overline{F}(z) \tag{6.2.37}$$

Differentiating both sides of (6.2.37) with respect to z and simplifying, we obtain

$$\psi(z)(r(z))^2 = g''(z) - 3g'(z)r(z) - g'(z)\frac{f'(z)}{f(z)} + g(z)(r(z))^2. \tag{6.2.38}$$

Using the relation

$$\frac{f'(z)}{f(z)} = \frac{r'(z)}{r(z)} - r(z), \tag{6.2.39}$$

Equation (6.2.38) can be written as

$$g'(z)\frac{r'(z)}{r(z)}+2g'(z)r(z)=g''(z)+(r(z))^2(g(z)-\psi(z))=0 \qquad (6.2.40)$$

Thus $r'(z)$ has been expressed in terms of functions of *r(z)* and the known functions. It can be shown that *r(z)* is the unique solution of the above equation. This completes the proof.

For the general case, the problem becomes more complicated because of the nature of the resulting differential equation.

Example 6.2.1.

If s= 1, ψ(x)=x and g(x) = ax+b, then from (6.2.35) we obtain

$$r(z)=\frac{\mathrm{a}}{(\mathrm{a}-1)z+b}. \qquad (6.2.41)$$

From we (5.2.41), we will obtain

i) F(x) = $1-e^{-ax/b}$, for a/b >0 and F(0) = 0,

ii) F(x)= $1-(x+\frac{\mathrm{b}}{\mathrm{a}-1})^{a/(a-1)}$, for a ≠1. Thus X will have power function distribution if a <1 and Pareto distribution if a>1.

Example 6.2.2.

If s=2 , ψ(x)=x and g(x) = ax+b, then from (6.2.40) we obtain

$$\mathrm{r(x)}=\frac{\mathrm{a}+\sqrt{\mathrm{a}}}{(\mathrm{a}-1)\,\mathrm{x}+\mathrm{b}}.$$

Thus X will have (i) exponentially distributed if a=1, (ii) power function distribution if a <1 and (iii) Pareto distribution if a >1.

Remark It will be interesting to consider the conditional expectation $E\left[\psi(R_{k+s})\middle|R_k=z\right]=g(z)$ *to* characterize distributions. In this case, the problem becomes more complicated because of the nature of the resulting differential equation.

Chapter 7

Characterizations Under the Assumption of Independence, Identical Distribution and Moment Properties

7.1. Characterizations under the Assumption of Independence

We have already seen that if $X \in E(0, \sigma)$, then $R_n - R_m$ and R_m , $n > m \geq 0$, are independent. This is a characteristic property of the exponential distribution. For n = 1, Tata (1969) proved the following characterization theorem.

Theorem 7.1.1.

Let $\{ X_n, n \geq 1\}$ be an i.i.d. sequence of non- negative random variables with cdf F(x) and pdf f(x). Then for X_n to belong to $E(\mu, \sigma)$, it is necessary and sufficient that R_1-R_0 and R_0 are independent.

Proof

The property of the independence of R_1-R_0 and R_0 will lead to the functional equation

$$\bar{F}(0)\bar{F}(x + y) = \bar{F}(x)\,\bar{F}(y),\ \ 0 < x, y < \infty . \tag{7.1.1}$$

The continuous solution of this functional equation with the boundary condition F(0) =0, is

$$\bar{F}(x) = e^{-x\sigma^{-1}}, \ x > 0, \sigma > 0.$$

The following theorem is a generalization of theorem 7.1.1.

Theorem 7.1.2.

Let $\{X_n, n \geq 1\}$ be a sequence of i.i.d. random variables with common distribution function F which is absolutely continuous with pdf f. Assume that F(0) =0. Then for $X_n \in E(0, \sigma)$, it is necessary and sufficient that R_{n-1} and $I_{n-1,n}$ $(= R_n - R_{n-1})$, $n \geq 1$, are independent.

Proof.

It is easy to establish that if $X_n \in E(0, \sigma)$, $n \geq 1$, then $I_{n-1,n}$ and R_{n-1} are independent. We will prove the sufficient condition. Suppose that R_{n-1} and $I_{n-1,n}$ are independent. Now the joint pdf f(z,u) of $I_{n-1,n}$ and R_{n-1} can be written as

$$f(z,u) = \frac{[R(u)]^{n-1}}{\Gamma(n)} r(u) f(u+z) \quad, 0 < u, z < \infty. \tag{7.1.2}$$

= 0, otherwise.

But the pdf $f_{n-1}(u)$ of R_{n-1} can be written as

$$F_{n-1}(u) = \frac{[R(u)]^{n-1}}{\Gamma(n)} f(u), 0 < u < \infty,$$

= 0, otherwise. (7.1.3)

Since $I_{n-1,n}$ and R_{n-1} are independent, we get from (7.1.2) and (7.1.3)

$$\frac{f(u+z)}{\bar{F}(u)} = g(z), \tag{7.1.4}$$

where g(z) is the pdf of $I_{n-1,n}$. Integrating (7.1.4) with respect z from 0 to z_1, we get

$$\bar{F}(u) - \bar{F}(u+z_1) = \bar{F}(u) G(z_1). \tag{7.1.5}$$

Since $G(z_1) = \int_0^{z_1} g(z)dz$. Now $u \to 0^+$ and using the boundary condition, we see that $G(z_1) = F(z_1)$. Hence we get from (7.1.5)

$$\bar{F}(u+z_1) = \bar{F}(u)\,\bar{F}(z_1). \tag{7.1.6}$$

The only continuous solution of (7.1.6) with the boundary condition F(0) = 0, is

$$\bar{F}(x) = e^{-\sigma^{-1}x}, x \geq 0 \tag{7.1.7}$$

where σ is an arbitrary positive real number.

Theorem 7.1.3 is a generalization of the Theorem 7.1.2.

Theorem 7.1.3.

Let $\{X_n, n \geq 1\}$ be independent and identically distributed with common distribution function F which is absolutely continuous and F(0) = 0 and F(x) < 1 for all x> 0. Then for $X_n \in E(0, \sigma)$, it is necessary and sufficient that $I_{m,n}$ (=$R_n - R_m$) and R_m ($n > m \geq 0$) are independent.

Proof.

The necessary condition is easy to establish. To proof the sufficient condition, we need the following lemma.

Lemma 7.1.1.

Let F(x) be an absolutely continuous function and $\bar{F}(x) > 0$, for all x > 0. Suppose that $\bar{F}(u+v)(\bar{F}(v))^{-1}$ = exp{-q(u,v)} and h(u,v) = $\{q(u,v)\}^r$ exp{-q(u,v)} $\frac{\partial}{\partial u}$ q(u,v), for $r \geq 0$. Further if h(u,v) ≠ 0, and $\frac{\partial}{\partial u}$ q(u,v) ≠ 0 for any positive u and v. If h(u,v) is independent of v, then q(u,v) is a function of u only.

Proof

Let

$$g(u) = h(u, v)$$

$$= (q(u,v))^r \exp(-q(u,v))\frac{\partial}{\partial v} q(u,v)$$

$$= \sum_{j=0}^{\infty} \frac{(-1)^j}{\Gamma(j+1)} \{q(u,v)\}^{r+j} \frac{\partial}{\partial u} q(u,v)$$

$$= \sum_{j=0}^{\infty} \frac{(-1)^j}{\Gamma(j+1)} \frac{1}{(r+j+1)} \frac{\partial}{\partial u} q(u,v).$$

Hence

$$\sum_{j=0}^{\infty} \frac{(-1)^j}{\Gamma(j+1)} (q(u,v))^{r+j+1} \frac{1}{(r+j+1)} = \mathrm{c} + \int g(u)du ,$$

$$= g_1(u), \text{say}. \tag{7.1.8}$$

Here $g_1(u)$ is a function of u only and c is independent of u but may depend on v. Now letting $u \to 0^+$, we see that $q(u,v) \to 0$ and hence from (7.1.8), we have c as independent of v. Therefore

$$0 = \frac{\partial}{\partial v} g_1(v) = \sum_{j=0}^{\infty} \frac{(-1)^j}{\Gamma(j+1)} \{q(u,v)\}^{r+j} \frac{\partial}{\partial v} q(u,v)$$

$$= g(u) \frac{\partial}{\partial v} q(u,v) (\frac{\partial}{\partial u} q(u,v))^{-1}.$$

Now we know $g(u) = h(u,v) \neq 0$ and $\frac{\partial}{\partial u} q(u,v) \neq 0$, so we must have $\frac{\partial}{\partial v} q(u,v) = 0$.

We prove the sufficiency condition of the Theorem 7.1.3. The joint pdf of $I_{m,n}$ and R_m is

$$f(z,u) = \frac{R^m(x) r(x)}{\Gamma(m+1)\Gamma(n-m)} \left[R(z+x) - R(x)\right]^{n-m-1} f(z+x), \ 0 < z < \infty, \ 0 < x < \infty.$$

The conditional pdf of $I_{m,n}$ given $R_m = x$ is

$$f(z|R_m = x) = \frac{1}{\Gamma(n-m)} \left[R(z+x) - R(x)\right]^{n-m-1} \frac{f(z+x)}{\overline{F}(x)}, \ 0 < z < \infty, \ 0 < x < \infty.$$

Since $I_{m,n}$ and R_m are independent, we will have for all z> 0,

$$(R(z+x)-R(x))^{n-m-1}\frac{f(z+x)}{\overline{F}(x)} \tag{7.1.9}$$

as independent of x. Now let

$$R(z+x)-R(x) = -\ln\frac{\overline{F}(z+x)}{\overline{F}(x)} = q(z,x), \text{ say.}$$

Writing (7.1.9) in terms of q(z,x), we get

$$[q(z,x)]^{n-m-1}\exp\{-q(z,x)\}\frac{\partial}{\partial z}q(z,x), \tag{7.1.10}$$

as independent of x. Hence by the lemma 7.1.1, we have

$$-\ln\{\overline{F}(z+x)(\overline{F}(x))^{-1}\} = q(z+x) = c(z), \tag{7.1.11}$$

where c(z) is a function of z only. Thus

$$\overline{F}(z+x)\,(\overline{F}(x))^{-1} = c_1(z), \tag{7.1.12}$$

and $c_1(z)$ is a function of z only.

The relation (7.1.12) is true for all $z \geq 0$ and any arbitrary fixed positive number x. The continuous solution of (7.1.12) with the boundary conditions, $\overline{F}(0)=1$ *and* $\overline{F}(\infty)=0$ is

$$\overline{F}(x) = \exp(-x\,\sigma^{-1}), \tag{7.1.13}$$

for $x \geq 0$ and any arbitrary positive real number σ. The assumption of absolute continuity of F(x) in the Theorem 7.1.3 can be replaced by the continuity of F(x).

7.2. Characterizations Based on Identical Distribution

We have seen that if the sequence $\{X_n,\ n \geq 1\}$ of i.i.d r.v.'s are from E(0,σ), then $R_{n-1} \overset{d}{=} Z_1+Z_2+\ldots+Z_n$, where $Z_1, Z_2,\ldots, Z_n$ are i.i.d. E(0, σ). The following theorem gives a characterization of the exponential distribution using the above property. If F is the distribution function of a non-negative random variable, we will call F is "new better than used" (NBU) if for x, y $\geq$ 0, $\overline{F}(x+y) \leq \overline{F}(x)\overline{F}(y)$, and F is "new worse than used"

(NWU) if for x, y ≥ 0, $\overline{F}(x+y) \geq \overline{F}(x)\overline{F}(y)$. We will say F belongs to the class C_1 if either F is NBU or NWU.

Theorem 7.2.1.

Let X_n, $n \geq 1$ be a sequence of i.i.d. random variables which has absolutely continuous distribution function F with pdf f and F(0) = 0. Assume that F(x) < 1 for all x>0. If X_n belongs to the class C_1 and $I_{n-1,n}$ has an identical distribution with X_k, $k \geq 1$, then $X_k \in E(0, \sigma)$, $k \geq 1$.

Proof.

The pdf g_n of $I_{n-1,n}$ can be written as

$$g_n(x) = \int_0^{\infty} \frac{[R(u)]^{n-1}}{\Gamma(n)} r(u) f(u+z)\, du\ , z \geq 0$$

$$= 0, \text{ otherwise.} \tag{7.2.1}$$

By the assumption of the identical distribution of $I_{n-1,n}$ and X_k, we must have

$$\int_0^{\infty} [R(u)]^{n-1} \frac{r(u)}{\Gamma(n)} f(u+z)du = f(z), \text{ for all } z > o. \tag{7.2.2}$$

Substituting

$$\int_0^{\infty} [R(u)]^{n-1} f(u)du = \Gamma(n), \tag{7.2.3}$$

we have

$$\int_0^{\infty} [R(u)]^{n-1} r(u) f(u+z)du = f(z) \int_0^{\infty} [R(u)]^{n-1} f(u)du\ , z > 0. \tag{7.2.4}$$

Thus

$$\int_0^{\infty} [R(u)]^{n-1} f(u)[f(u+z)(\overline{F}(u))^{-1} - f(z)]\, du = 0, z > 0. \tag{7.2.5}$$

Integrating the above expression with respect to z from z_1 to ∞, we get from (7.2.5)

$$\int_0^\infty [R(u)]^{n-1} f(u)\left[\bar{F}(u+z_1)(\bar{F}(u))^{-1} - \bar{F}(z_1)\right] du = 0,\ z_1 > 0. \qquad (7.2.6)$$

If F(x) is NBU, then (7.2.6) is true if

$$\bar{F}(u+z_1)(\bar{F}(u))^{-1} = \bar{F}(z_1),\ z_1 > 0. \qquad (7.2.7)$$

The only continuous solution of (7.2.7) with the boundary conditions $\bar{F}(0) = 1$ and $\bar{F}(\infty) = 0$ is $\bar{F}(x) = \exp(0, \sigma^{-1})$, where σ is an arbitrary real positive number. Similarly, if F is NWU then (7.2.6) is true if (7.2.7) is satisfied and hence $X_k \in E(0, \sigma)$, $k \geq 1$.

The following theorem is proved under the assumption of monotone hazard rate. We will say F belongs to the class C_2 if r(x) is either monotone increasing or decreasing.

Theorem 7.2.2.

If X_k, $k \geq 1$ has an absolutely continuous distribution function F with pdf f and F(0) = 0. If $I_{n,n+1}$ and $I_{n-1,n}$, $n \geq 1$, are identically distributed and F belongs to C_2, then $X_k \in E(0,\sigma)$, $k \geq 1$.

Proof

$$P(I_{n-1,n} > z) = \int_0^\infty [R(u)]^{n-1} \frac{r(u)}{\Gamma(n)} \bar{F}(u+z)du, \text{ for all } z > 0,$$

= 0, otherwise.

Since $I_{n,n+1}$ and $I_{n-1,n}$ are identically distributed, we get using the above equation,

$$\int_0^\infty [R(u)]^n r(u)\bar{F}(u+z)du = n\int_0^\infty [R(u)]^{n-1} r(u)\bar{F}(u+z)du,\ z > 0. \qquad (7.2.8)$$

Substituting the identity

$$n\int_0^\infty [R(u)]^{n-1} r(u)\bar{F}(u+z)du = \int_0^\infty [R(u)]^n f(u+z)du$$

in (7.2.8), we get on simplification

$$\int_0^\infty [R(u)]^{n-1} r(u)\bar{F}(u+z)\left[1-\frac{r(u+z)}{r(u)}\right]du = 0, \ z>0. \qquad (7.2.9)$$

Thus if $F \in C_2$, then (7.2.9) is true if for almost all u and any fixed $z > 0$,

$$r(u+z) = r(u) . \qquad (7.2.10)$$

The constant hazard rate i.e. the relation (7.2.10) is a well known characterization property of the exponential distribution. Hence, we have $X_k \in E(0, \sigma)$ for all $k \geq 1$.

Theorem 7.2.3.

Let X_n, $n \geq 1$ be a sequence of independent and identically distributed non-negative random variables with absolutely continuous distribution function F(x) with f(x) as the corresponding density function. If $F \in C_2$ and for some fixed n,m, $1 \leq m < n < \infty$, $I_{m,n} \stackrel{d}{=} R_{n-m-1}$, then

$X_k \in E(0,\sigma)$, $k \geq 1$.

Proof.

The pdfs $f_1(x)$ of R_{n-m} and $f_2(x)$ of $I_{m,n}$ ($= R_n - R_m$) can be written as

$$f_1(x) = \frac{1}{\Gamma(n-m)}[R(x)]^{n-m-1} f(x) \text{ , for } 0 < x < \infty, \qquad (7.2.11)$$

and

$$f_2(x) = \int_0^\infty \frac{[R(u)]^{m-1}}{\Gamma(m)} \frac{[R(x+u)-R(x)]^{n-m-1}}{\Gamma(n-m)} r(u) f(u+x)du \text{, } 0 < x < \infty \qquad (7.2.12)$$

Integrating (7.2.11) and (7.2.12) with respect to x from o to x_0, we get

$$F_1(x_0) = 1 - g_1(x_0), \qquad (7.2.13)$$

where

$$g_1(x_0)=\sum_{j=1}^{n-m}\frac{[R(x_0)]^{j-1}}{\Gamma(j)}e^{-R(x_o)},$$

and

$$F_2(x_0) = 1 - g_2(x_0, u), \tag{7.2.14}$$

where

$$g_2(x_0,u)=\sum_{j=1}^{n-m}\frac{[R(u+x_o)-R(u)]^{j-1}}{\Gamma(j)}\exp\{-(R(u+x_o)-R(u))\}.$$

Now equating (7.2.13) and (7.2.14), we get

$$\int_0^\infty \frac{[R(y)]^{m-1}}{\Gamma(m)}f(u)[g_2(u,x_0)-g_1(x_0)]\,du=0,\ x_0>0. \tag{7.2.15}$$

Now $g_2(x_0, 0) = g_1(0)$ and

$$0=\frac{[R(u)-R(u)]^{n-m-1}}{\Gamma(n-m)}\exp\{-(R(u+x_o)-R(u)\}[r(x_o)-r(u+x_o)].$$

Thus if $F \in C_2$, then (7.2.15) is true if

$$r(u+x_0) = r(u) \tag{7.2.16}$$

for almost all u and any fixed $x_0 \geq 0$. Hence $X_k \in E(0,\sigma)$ for all $k \geq 1$. Here σ is an arbitrary positive real number. Substituting m = n-1, we get $I_{n-1,n} \stackrel{d}{=} X_1$ as a characteristic property of the exponential distribution.

Theorem 7.2.4.

Let $\{X_n, n \geq 1\}$ be a sequence of independent and identically distributed non-negative random variables with absolutely continuous distribution function F(x) and the corresponding density function f(x). If F belongs to C_2 and for some m, $m > 1$, R_m and $R_{m-1} + U$ are identically distributed, where U is independent of R_m and R_{m-1} and is distributed as X_n's, then

$X_n \in E(0, \sigma)$, for some $\sigma > 0$.

Proof

The pdf $f_m(x)$ of R_m, $m \geq 1$, can be written as

$$f_m(y) = \frac{[R(y)]^m}{\Gamma(m+1)} f(y),\ 0<y<\infty,$$

$$= \frac{d}{dy}\left(-\bar{F}(y) \int_0^y \frac{[R(x)]^{m-1}}{\Gamma(m)} r(x)\, dx + \int_0^y \frac{[R(x)]^m}{\Gamma(m)} f(x) dx \right), \tag{7.2.17}$$

The pdf $f_2(y)$ of $R_{m-1} + U$ can be written as

$$f_2(y) = \int_0^y \frac{[R(x)]^{m-1}}{\Gamma(m)} f(y-x) f(x)\, dy$$

$$= \frac{d}{dy}\left(-\frac{[R(x)]^{m-1}}{\Gamma(m)} \bar{F}(y-x) f(x)\, dx + \int_0^y \frac{[R(x)]^{m-1}}{\Gamma(m)} f(x)\, dx \right). \tag{7.2.18}$$

Equating (7.2.17) and (7.2.18), we get on simplification

$$\int_0^y \frac{[R(x)]^{m-1}}{\Gamma(m-1)} f(x) H_1(x,y)\, dx = 0, \tag{7.2.19}$$

where $H_1(x,y) = \bar{F}(y-x) - \bar{F}(y)(\bar{F}(x))^{-1}$, $0<x<y<\infty$. Since $F \in C_1$, therefore for (7.2.19) to be true , we must have

$$H_1(x,y) = 0, \tag{7.2.20}$$

for almost all x, $0 < x < y < \infty$.

This implies that

$$\bar{F}(y-x)\,\bar{F}(x) = \bar{F}(y), \tag{7.2.21}$$

for almost all x, $0 < x < y < \infty$. The only continuous solution of (7.2.21) with the boundary conditions $\bar{F}(0) = 1$, *and* $\bar{F}(\infty) = 0$, *is*

$$\overline{F}(x) = e^{-x\sigma^{-1}}, \tag{7.2.22}$$

where σ is an arbitrary positive number.

Remark 7.2.1.

The theorem 7.7 can be used to obtain the following known results of a two parameter exponential distribution($\overline{F}(x) = \exp\{-\sigma^{-1}(x-\mu)\}$).

$E(R_{m-1}) = \mu + m\,\sigma$

$Var(R_{m-1}) = m\,\sigma^2$

$Cov(R_{m-1}, R_{n-1}) = m\,\sigma^2$, $m < n$.

Theorem 7.2.5.

Let $X_1, X_2, \ldots, X_m, \ldots$ be independent and identically distributed random variables with probability density function f(x), $x \geq 0$ and m is an integer valued random variable independent of X's and $P(m = k) = p(1-p)^{k-1}$, k = 1, 2, ..., and $0 < p < 1$. Then the following two properties are equivalent:

(a) X's are distributed as $E(0,\sigma)$, where σ is a positive real number

(b) $p \sum_{j=1}^{m} X_j \stackrel{d}{=} I_{n-1,n}$, for some fixed n, $n \geq 2$, $X_j \in c_2$ and $X_j \in c_2$ and $E(X_j) < \infty$.

Proof.

It is easy to verify (a) ⇒ (b). We will prove here that (b) ⇒ (a). Let $\phi_1(t)$ be the characteristic function of $I_{n-1,n}$ then

$$\phi_1(t) = \int_0^\infty \int_0^\infty \frac{1}{\Gamma(n)} e^{itx} [R(u)]^{n-1} r(u) f(u+x)\, du\, dx$$

$$= 1 + it \int_0^\infty \int_0^\infty \frac{1}{\Gamma(n)} e^{itx} [R(u)]^{n-1} r(u) \overline{F}(u+x)\, du\, dx$$

can be written as

$$\Phi_2(t) = E(e^{itp\sum_{j=1}^{m} X_j})$$

$$= \sum_{k=1}^{\infty} [\Phi(tp)]^k p(1-p)^{k-1}, \tag{7.2.23}$$

$$= p(\Phi(t\,p))\,(1 - q\,\Phi(p\,t))^{-1},\ q = 1-p, \tag{7.2.24}$$

where Φ(t) is the characteristic function of X's. Equating (7.2.23) and (7.2.24), we get on simplification

$$\frac{\Phi(pt)-1}{1-q\Phi(pt)}\frac{1}{it} = \int_0^{\infty}\int_0^{\infty} \frac{1}{\Gamma(n)} e^{itx} [R(u)]^{n-1} r(u) \bar{F}(u+x)\, du dx, \tag{7.2.25}$$

Now taking limit of both sides of (7.2.25) as t goes to zero, we have

$$\frac{\Phi'(0)}{i} = \int_0^{\infty}\int_0^{\infty} \frac{1}{\Gamma(n)} [R(u)]^{n-1} r(u) \bar{F}(u+x) du dx\ . \tag{7.2.26}$$

Writing

$$\frac{\Phi'(0)}{i} = \int_0^{\infty} \bar{F}(x)dx,\ we\ get\ from (2.4.26),$$

$$\int_0^{\infty}\int_0^{\infty} (R(u))^{n-1} r(u) \{\bar{F}(u+x) - \bar{F}(u)\bar{F}(x)\} du\, dx = 0. \tag{7.2.27}$$

Since X's belong to C_1, we must have

$$\bar{F}(u+x) = \bar{F}(x)\bar{F}(u)\ , \tag{7.2.28}$$

for almost all x, u, $0 < u, x < \infty$. The only continuous solution of (7.2.28) with the boundary condition $\bar{F}(0) = 1\ and\ \bar{F}(\infty) = 0$, is

$$\bar{F}(x) = \exp(-x\,\sigma^{-1}), x \ge 0\ , \tag{7.2.29}$$

where σ is an arbitrary positive real number.

We will prove the following characterization theorem under the assumption of the finite first moment.

7.3. CHARACTERIZATION OF EXPONENTIAL DISTRIBUTION

Theorem 7.3.1.

Let X_n, $n \geq 1$ be a sequence of independent and identically distributed non-negative random variables with absolutely continuous distribution function F(x) and the corresponding density function f(x). Let a = inf{x|F(x) >0} =0 , F(x) < 1 for all x > 0. If F belongs to the class C_1 and $E(X_k)$, $k \geq 1$ is finite, then $X_k \in E(0,\sigma)$, if and only if for some fixed n, n > 1, $E(I_{n-1,n}) = E(X_k)$.

Proof.

If $X_k \in E(0,\sigma)$, then it can easily be seen that $E(I_{n-1,n}) = E(X_k)$. Suppose that for some foxed n, n> 1, $E(I_{n-1,n}) = E(X_k)$, then we must have

$$\int_0^\infty \int_0^\infty \frac{[R(u)]^{n-1}}{\Gamma(n)} f(u)(\bar{F}(u))^{-1}\, du dx = \int_0^\infty \bar{F}(u) du \tag{7.3.1}$$

But we know

$$\Gamma(n) = \int_0^\infty [R(u)]^{n-1} f(u) du \tag{7.3.2}$$

Substituting (7.3.2) in (7.3.3) and simplifying, we have

$$\bar{F}(u+z) = \bar{F}(u)\bar{F}(z), \tag{7.3.3}$$

for all u,z, 0 < u,z < ∞. Now the continuous solution of (7.3.3) with the boundary conditions $\bar{F}(0) = 1$ and $F(\infty) = 0$, is $\bar{F}(x) = \exp\{-x\sigma^{-1}\})$, where σ is an arbitrary real number.

The following theorem uses the property of homocedasticity but does not use NBU or NWU property.

Theorem 7.3.2.

Let X_n, $n \geq 1$ be a sequence of independent and identically distributed random variables with common distribution function F which is absolutely continuous and inf{x| F(x) > 0} =0 and $E(X_n^2) < \infty$. Then X_k, $k \geq 1$ has the exponential distribution if and only if $Var(I_{n-1,n} | R_{n-1} = x) = b$ for all x, where b is a positive constant independent of X and $I_{n-1,n}$.

Proof.

The 'if" condition is easy to establish. We will prove here the ' only if ' condition. Now

$$b = E(I_{n-1,n}^{2} \mid R_{n-1} = x) - [E(I_{n-1,n} \mid R_{n-1} = x)]^2. \quad (7.3.4)$$

Also

$$E(I_{n-1,n}^{2} \mid R_{n-1} = x) = \int_0^\infty z^2[\overline{F}(x)]^{-1}\, d\overline{F}(z+x) = 2\int_0^\infty z[\overline{F}(x)]^{-1}\,\overline{F}(z+x)dz \quad (7.3.5)$$

and

$$E(I_{n-1,n} \mid R_n = x) = \int_0^\infty z(\overline{F}(x))^{-1}\, d\overline{F}(z+x) = \int_0^\infty (\overline{F}(x))^{-1}\,\overline{F}(z+x)\, dz \quad (7.3.6)$$

Substituting $G(x) = \int_0^\infty z\ \overline{F}(z+x)dz$ and denoting $G^{(r)}(x)$ as the rth derivative of G(x), we have on simplification

$$G^{(1)}(x) = \int_0^\infty \overline{F}(z+x)dz,\ G^{(2)}(x) = \overline{F}(x)\ and\ G^{(3)}(x) = -f(x).$$

Writing (7.3.5) and (7.3.6) in terms of G(x) and $G^{(r)}(x)$, we get from (7.3.3)

$$2G(x)\,\{G^{(r)}(x)\}^{-1} - \{G^{(1)}(x)\,(G^{(2)}(x))^{-1}\}^2 = b, \text{ for all } x > 0. \quad (7.3.7)$$

Differentiating (7.3.7) with respect to x and simplifying, we obtain

$$2G^{(3)}(x)\,\{G^{(2)}(x)\}^{-3} - [\,(G^{(1)}(x))^2 - G(x)\,G^{(2)}(x)] = 0 \quad (7.3.8)$$

Since $G^{(3)}(x) \neq 0$ for all $x > 0$, we must have

$$\{G^{(1)}(x)\}^2 - G(x)\,G^{(2)}(x) = 0, \quad (7.3.9)$$

i.e.

$$\frac{d}{dx}\{G(x)\,(G^{(1)}(x))^{-1}\} = 0, \text{ for all } x > 0. \quad (7.3.10)$$

The solution of (7.3.10) is

$$G(x) = a\,e^{-cx},\ x > 0 \quad (7.3.11)$$

where a and c are arbitrary constants. Hence

$$\bar{F}(x) = G^{(2)}(x) = ac^2 e^{-cx}, \; x>0.$$

Since F(x) is a distribution function with F(0) = 0, it follows that

$$\bar{F}(x) = e^{-x\sigma^{-1}},$$

where σ is an arbitrary real positive number. The following Theorem gives a characterization of the exponential distribution using the hazard rate.

Theorem 7.3.3.

Let $\{X_n, n \geq 1\}$ be a sequence of independent and identically distributed non negative random variables with continuous distribution function F(x) and the corresponding density function f(x). Let $a = \inf\{x|F(x) = 0\} = 0$, $F(x) < 1$ for all $x > 0$ and F belongs to class C_2. Then $X_k \in E(0,\sigma)$, if and only if for some fixed n, $n \geq 1$, the hazard rate r_1 of $I_{n-1,n}$= the hazard rate r of X_k.

Proof.

If $X_k \in E(0,\sigma)$, then it can easily be shown that $r_1 = r$. Suppose $r_1 = r$, then we can write the joint pdf of R_n and R_{n-1} as

$$f_{n-1,n}(x,y) = \frac{1}{\Gamma(n)}[R(x)]^{n-1} r(x) f(y), \; 0 < x < y < \infty,$$

= 0, otherwise.

Substituting $I_{n-1,n} = R_n - R_{n-1}$ and $U = R_{n-1}$, we get the pdf of $I_{n-1,n}$ and U as

$$f_1^*(z,u) = \frac{1}{\Gamma(n)}[R(u)]^{n-1} r(u) f(u+z), \; 0 < u, z < \infty, \qquad (7.3.12)$$

= 0, otherwise.

Thus by (7.3.12), we can write

$$r_1(z) = \frac{\int_0^\infty [R(u)]^{n-1} r(u) f(u+z) du}{\int_0^\infty [R(u)]^{n-1} r(u) \bar{F}(u+z) du}. \tag{7.3.13}$$

for all $z \geq 0$. Since $r_1(z) = r(z)$ for all z, we must have

$$\frac{\int_0^\infty [R(u)]^{n-1} r(u) f(u+z) du}{\int_0^\infty [R(u)]^{n-1} r(u) \bar{F}(u+z) du} = \frac{f(z)}{\bar{F}(z)} \tag{7.3.14}$$

for all $z \geq 0$. Now simplifying (7.3.14), we have

$$\int_0^\infty [R(u)]^{n-1} r(u) \bar{F}(z) \bar{F}(u+z)[r(u+z) - r(z)]\, du = 0 \tag{7.3.15}$$

for all $z \geq 0$. Since F belongs to class C_2 , for (7.3.15) to be true, we must have

$$r(u+z) = r(u) \tag{7.3.16}$$

for all $z \geq 0$ and almost al u, $u \geq 0$. Hence $X_k \in E(0, \sigma)$.

The exponential distribution can also be characterized using lower record values. The following result is due to Ahsanullah and Kirmani (1991).

Lemma 7.3.1.

Suppose $\{X_n, n \geq 1\}$ be a sequence of i.i.d. random variables with cdf F and $F(0) = 0$. Let N is the r.v. defined as $N = \min\{ i > 1: X_i < X_1\}$. It can easily be shown that $P(N = n) = \frac{1}{n(n-1)}$, $n = 2,3,\ldots,$.

Theorem 7.3.4.

The rvs NX_N and X_1 are identically distributed iff $F(x) = \exp(-\lambda x)$, $x > 0$, for some $\lambda > 0$.

In proving the theorem , we need the following Lemma.

Lemma 7.3.2.

$$P(NX_N>x)=\sum_{n=2}^{\infty} \frac{1}{n(n-1)} (\bar{F}(\frac{x}{n}))^n, \text{ for all } x \geq 0.$$

Proof.

$$P(NX_N>x)=\sum_{n=2}^{\infty} P(NX_N>x, N>n)$$

$$= \sum_{n=2}^{\infty} \int_{-\infty}^{\infty} P(nX_n> x, X_n < y, X_i>y \mid X_1 = y)\, dF(y), \text{ for all } i = 2,3,\ldots, n-1,$$

$$= \sum_{n=2}^{\infty} \int_{\frac{x}{n}}^{\infty} P(\frac{x}{n} <x_n< y)(P(x_i>y))^{n-2} dF(y)$$

$$= \sum_{n=2}^{\infty} \int_{\frac{x}{n}}^{\infty} \{\bar{F}(\tfrac{x}{n})-\bar{F}(y)\}\{\bar{F}(y)\}^{n-2} dF(y)$$

$$= \sum_{n=2}^{\infty} \left[\bar{F}(\tfrac{x}{n})\frac{[\bar{F}(x)]^{n-1}}{n-1} - \frac{[\bar{F}(\frac{x}{n})]^n}{n}\right]$$

$$= \sum_{n=2}^{\infty} \frac{1}{n(n-1)}[\bar{F}(\tfrac{x}{n})]^n.$$

Proof of the Theorem 7.3.4.

Define $u(x) = -\dfrac{\ln \bar{F}(x)}{x}$, $x > 0$; $u(0) = u(0+)$ and suppose that $NX_N \stackrel{d}{=} X_1$. Then

$$\sum_{n=2}^{\infty} \frac{1}{n(n-1)} e^{-xu(x/n)} = e^{-xu(x)}, \ x > 0. \tag{7.3.17}$$

We shall show that the above holds iff u(x) is a constant, i.e. given any T > 0

$$\min_{x\in[0,T]} U(x)=\max_{x\in[0,T]} U(x). \tag{7.3.18}$$

Let

$$a_0=\min_{x\in[0,T]} U(x),\ x_0=\inf\{x\in[0,T]\,|u(x)=a_0\},$$

$$a_1=\max_{x\in[0,T]} U(x),\ x_1=\inf\{x\in[0,T]\,|\,u(x)=a_1\}.$$

It is obvious that (7.3.18) will be proved if we show that $x_0 = 0 = x_1$. By continuity of u, $x_0 \in[0,T]$ and u(x_0)= a_0. Hence

$$u(x_0)\le u(x_0\ /\ n) \text{ for all n} \geq 1. \tag{7.3.19}$$

If equality holds for all n $\geq$ 2, then $u(x_0)=u(0)$ which by definition of x_0 = 0. Suppose now that x_0 > 0 (so that x_0 / n $\neq$ x_0 for all n. Then, the strict inequality must hold for (7.3.19) for at least one value of n >1 $\geq$ 2. Now

$$e^{-x_0u(x_0)}-\sum_{n=2}^{\infty}\frac{1}{n(n-1)}e^{-x_0u(x_0/n)}=\sum_{n=2}^{\infty}\frac{1}{n(n-1)}\{e^{-x_0u(x_0)}-e^{-x_0u(x_o/n)}\}>0,$$

which contradicts (7.3.19). Therefore x_0 = 0. Similarly x_1 = 0. Thus $NX_N \overset{d}{=} X_1 \Rightarrow u(x) \equiv$ constant. The converse can easily be verified.

Basak(1996) give a similar characterization based on k-records. The result is given in the following theorem.

Theorem 7.3.5.

Suppose {X_n, n $\geq$1} be a sequence of i.i.d. random variables with cdf F with F(0) = 0 and $\lim_{x\to 0^+}\frac{F(x)}{x}=\lambda$, λ >0. If ($T_{(n)}$(k) –k+1) $R_{(n)}$(k) and $X_{1,n}$, k $\geq$ 1, are identically distributed, then X has the exponential distribution with F(x) = exp (-λx).

Proof.

For any -∞ <x_n< x_{n-1}<.. x_1 < ∞ and d(0) = 0, k = d(1) < d(2) <... < d(n), we have

P($R_{(1)}$(k) $\leq$ x_1, $R_{(2)}$(k) $\leq$ x_2, ...,$R_{(n)}$(k) $\leq$ x_n, $T_{(1)}$(k) =k, $T_{(2)}$(k) =d(2),..., $T_{(n)}$(k) = d(n))

$$= \int_{-\infty}^{x_1} \cdots \int_{-\infty}^{x_n} \{\prod_{r=1}^{n} (\overline{G}(w_r))^{d(r)-d(r-1)-1} g(w_1) g(w_2) .... g(w_n) \; dw_1 \, dw_2 \ldots dw_n, \quad (7.3.20)$$

where

$$\overline{G}(x) = (\overline{F}(x))^k \text{ and } g(x) = -\frac{d}{dx}\overline{G}(x).$$

The joint density function, $f_{(1,2,...,n),(1),(2),..(n)}(x_1,x_2,...x_n,d(1),d(2),..d(n))$, of the continuous random variables $R_{(1)}(k)$, $R_{(2)}(k)$, …, $R_{(n)}(k)$ and the discrete record times $T_{(1)}(k)$, $T_{(2)}(k)$, …, $R_{(n)}(k)$ is obtained by differentiating the above equation with respect to $x_1,x_2,..x_n$. Thus

$f_{(1,2,...,n),(1),(2),...,(n)}(x_1,x_2,...x_n,d(1),d(2),..d(n))$

$$= \prod_{r=1}^{n} (\overline{G}(w_r))^{d(r)-d(r-1)-1} g(w_1) g(w_2) .... g(w_n) \,, \quad (7.3.21)$$

for $-\infty < x_n < x_{n-1} < .. \; x_1 < \infty$ and $d(0) = 0$, $k = d(1) < d(2) < ... < d(n)$.

On integrating out and simplification, we obtain from (7.3.21) the joint density function of $R_{(n)}(k)$ and $T_{(n)}(k)$ as

$$f(x, m_n) = \frac{|S(m_n - 1, n-1)|}{\Gamma(m_n - k + 2)} (\overline{G}(x))^{m_n - k} g(x), \; m_n \geq n-k+1. \quad (7.3.22)$$

Let $U = (T_{(n)}(k) - k + 1) R_{(n)}(k)$ and $M = T_{(n)}(k) - k + 1$, then the probability density of U is

$$f_U(x) = \sum_{m=n}^{\infty} \frac{|S(m-1, n-1)|}{\Gamma(m+1)} [\overline{G}(x/m)]^{m-1} g(x/m), \; m \geq n. \quad (7.3.23)$$

On integrating (7.3.23) from x to ∞, we obtain

$$1 - F_U(x) = \sum_{m=n}^{\infty} \frac{|S(m-1, n-1)|}{\Gamma(m+1)} [\overline{G}(x/m)]^m, \; x \geq 0, \; m \geq n. \quad (7.3.24)$$

Since $U = (T_{(n)}(k) - k + 1) R_{(n)}(k)$ and $X_{1,k}$, $k \geq 1$, are identically distributed, we must have

$$\sum_{m=n}^{\infty} \frac{|S(m-1, n-1)|}{\Gamma(m+1)} (\overline{G}(x/m))^m = \overline{G}(x), \text{ for all } x \geq 0. \quad (7.3.25)$$

The unique solution of the equation (7.3.25) with the conditions F(0) = 0 and $\lim_{x \to 0^+} \frac{F(x)}{x} = \lambda$ is

$$\overline{G}(x) = e^{-\lambda x / k}, \lambda > 0.\ k \geq 1 \text{ and } x \geq 0.$$

The following Theorem replaces the equality of distribution in Theorem 7.3.5 by the equality of expectation.

Theorem 7.3.6.

Suppose $\{X_n, n \geq 1\}$ be a sequence of i.i.d. random variables with cdf F with F(0) = 0. and $\lim_{x \to 0^+} \frac{F(x)}{x} = \lambda$, $\lambda > 0$. Further assume that $E(X_n) < \infty$. If

$$E((T_{(n)}(k) - k + 1)\, R_{(n)}(k)) = E(X_{1,n})\ k \geq 1,$$

then X_1 has the exponential distribution with $F(x) = \exp(-\lambda x)$.

Proof.

$E\,(T_{(n)}(k) - k + 1)\, R_{(n)}(k)) =$

$$\int_0^\infty \sum_{m=n}^{\infty} \frac{|S(m-1, n-1)|}{\Gamma(m+1)} [\overline{G}(x/m)]^m \,\text{dx, by} \qquad (7.3.25),$$

$$= \sum_{m=n}^{\infty} \frac{|S(m-1, n-1)|}{\Gamma(m+1)} \int_0^\infty [\overline{G}(x/m)]^m \,\text{dx} \qquad (7.3.26)$$

Now

$$E(X_{1,k)}) = \int_0^\infty (\overline{G}(x))\, dx$$

$$= \sum_{m=n}^{\infty} \frac{|S(m-1, n-1)|}{\Gamma(m+1)} \int_0^\infty (\overline{G}(x))\, dx \qquad (7.3.27)$$

Equating (7.3.26) and (7.3.27) and the proof of Theorem 7.3.5, the result can easily be established.

7.4. Characterization of Gumbel Distribution

If $\{X_j, j=1,\ldots,\}$ are from the Gumbel distribution with cdf as given in (1.2.2), then the pdf $f_{(m)}(x)$ of $X_{L(m)}$ is

$$f_{(m)}(x)=\frac{e^{-mx}}{\Gamma(m+1)}e^{-e^{-x}}, -\infty<x<\infty .$$

It is known that (see Ahsanullah and Holland (1994), p. 475) that

$$R_{(m)} \stackrel{d}{=} X-(W_1+\frac{W_2}{2}+\ldots+\frac{W_{m-1}}{m-1}+\frac{W_m}{m}),\ m\geq 1$$

where $W_1, W_2, \ldots, W_{m-1}, W_m$ are independently distributed as negative exponential with $F(w) = 1- e^{-w}$, $w>0$. $R_{(0)}=X$. Thus $S_{(m)} = m(R_{(m-1)} - R_{(m)})$, $m=1,2,\ldots$, are identically distributed as negative exponential. Similarly if we consider the upper records from the distribution, $F(x) = e^{-e^{x}}$, $-\infty < x < \infty$, , then for any $m \geq 1$, $S_m = m(R_{(m)} - R_{(m-1)})$ are identically distributed as negative exponential distribution. It can be shown that for one fixed m, $S_{(m)}$ or S_m distributed as negative exponential does not characterize the exponential distribution.

Arnold and Villasenor (1997) raised the question suppose that S_1 and 2 S_2 are i.i.d. negative exponential with unit mean, can we consider that X_j's are (possibly translated) Gumbel variables? Here, we will prove that for a fixed m >1, the condition $R_{(m)} \stackrel{d}{=} R_{(m+1)}+\frac{W}{m}$, where W is distributed as negative exponential distribution with mean unity characterizes the Gumbel distribution.

Theorem 7.4.1.

Let $\{X_j, j =1,\ldots,\}$ be a sequence of independent and identically distributed random variables with absolutely continuous (with respect to Lebesgue measure) distribution function F(x). Then the following two statements are identical.

(a) $F(x) = e^{-e^{-x}}$, $-\infty<x<\infty$,

For a fixed m >1, the condition $R_{(m)} \stackrel{d}{=} R_{(m+1)}+\frac{W}{m}$, where W is distributed as negative exponential mean unity.

Proof.

It is enough to show that (b) ⇒ (a). Suppose that for a fixed m > 1, $R_{(m)} \stackrel{d}{=} R_{(m+1)} + \frac{W}{m}$, then

$$F_{(m)}(x) = \int_{-\infty}^{x} P(W \le m(x-y) f_{(m+1)}(y)\,dy$$

$$= \int_{-\infty}^{x} [1 - e^{-m(x-y)}] f_{(m+1)}(y)\,dy$$

$$= F_{(m+1)}(x) - \int_{-\infty}^{x} e^{-m(x-y)} f_{(m+1)}(y)\,dy. \tag{7.4.1}$$

Thus

$$e^{mx}[F_{(m+1)}(x) - F_{(m)}(x)] = \int_{-\infty}^{x} e^{my} f_{(m+1)}(y)\,dy \tag{7.4.2}$$

Using the relation

$$e^{mx} \frac{F(x)[H(x)]^m}{\Gamma(m+1)} = e^{H(x)} \sum_{j=0}^{m} \frac{[H(x)]^j}{m!},$$

we obtain

$$e^{mx} \frac{F(x)(H(x))^m}{\Gamma(m+1)} = \int_{-\infty}^{x} e^{my} f_{(m+1)}(y)\,dy \tag{7.4.3}$$

Taking the derivatives of both sides of (7.4.3), we obtain

$$\frac{d}{dx}\left[e^{mx} \frac{(H(x))^m}{\Gamma(m+1)} F(x) \right] = e^{mx} f_{(m+1)}(x) \tag{7.3.4}$$

This implies that

$$\frac{d}{dx}\left[e^{mx} \frac{H^m(x)}{\Gamma(m+1)} \right] F(x) = 0\,. \tag{7.4.5}$$

Thus

$$\frac{d}{dx}\left[e^{mx}\frac{(H(x))^m}{\Gamma(m+1)}\right]=0\ . \tag{7.4.6}$$

Hence

$$H(x) = c\ e^{-x},\ -\infty < x < \infty \tag{7.4.7}$$

Thus

$$F(x)=\ e^{-ce^{-x}},\ \infty < x < \infty\ . \tag{7.4.8}$$

Since F(x) is a distribution function we must have c as positive. Assuming $F(0) = e^{-1}$, we obtain

$$F(x) = e^{-e^{-x}},\ -\infty < x < \infty. \tag{7.4.9}$$

In fact, $R_{(n)} \overset{d}{=} R_{(n-1)} + W$, will give a characterization of the exponential distribution.

Remark 7.4.1.

The conditions that any one of the statistics $m(R_{(m)} - R_{(m+1)})$, $m\,(R_{m+1} - R_m)$, $(R_{(m)} - R_{(m+1)})$ or $(R_{m+1} - R_m)$ is distributed as negative exponential do not characterize any distribution.

Recently Ahsanullah and Malov (2004) proved the following characterization theorem.

Theorem 7.4.2.

Let X_1, X_2, ..., be a sequence of independent and identically distributed r.v.'s with distribution function F(x). If $R_{(m)} \overset{d}{=} R_{(m+2)} + \frac{W_1}{m} + \frac{W_2}{m+1}$, for twice differentiable F(x), where W_1 and W_2 are independent as exponential distribution with unit mean then $F(x) = 1 - e^{-e^{-x}}$, $-\infty < x < \infty$.

Chapter 8

CHARACTERIZATIONS OF DISCRETE DISTRIBUTIONS

8.1. DEFINITION OF RECORDS OF DISCRETE DISTRIBUTIONS

Let $X_1, X_2, \ldots, X_n, \ldots$ be a sequence independent and identically distributed random variables taking values on 0,1,2,…such that F(n) <1 for all n n=0,1,2,…,. We define the upper record times, $T_{d,n}$ as $T_{d,0} = 1$, $T_{d,n+1} = \min\{j > T_{d,n}, X_j > X_{T_n}\}$, n= 0, 1, 2,…,. The nth upper record value is defined as $R_{d,n} = X_{T_{d,n}}$. Let $p_k = P(X_1 = k)$, $P(k) = \sum_{j=0}^{k} p(j)$, $k \geq 0$ and $\overline{P}(k) = 1 - P(k)$ with $P(\infty) = 1$. The joint probability mass function (pmf) of $R_{d,0}, R_{d,1}, \ldots, R_{d(n)}$ is defined as

$$P_{0,1,2,\ldots,n}(x_0, x_1, \ldots, x_n) = P(R_{d,0} = x_0, R_{d,1} = x_1, \ldots, R_{d,n} = x_n)$$

$$= \frac{p(x_0)}{\overline{P}(x_0)} \frac{p(x_1)}{\overline{P}(x_1)} \cdot \frac{p(x_2)}{\overline{P}(x_2)} \cdots \frac{p(x_{n-1})}{\overline{P}(x_{n-1})} p(x_n), \quad 0 \leq x_0 < x_1 < x_2 < \ldots < x_n < \infty$$

$$= 0, \text{ otherwise.} \tag{8.1.1}$$

The marginal pmf's of the upper record values are given as

$$P_{d,0}(x_0) = P(R_{d,0} = x_0) = p(x_0), \; x_0 = 0, 1, 2, \ldots,$$

$$P_{d,1}(x_1) = P(R_{d,1} = x_1) = S_1(x_1)\, p(x_1),$$

where

$$S_1(x_1) = \sum_{0 \leq x_0 < x_1} B(x_1), \quad B(x) = \frac{p(x)}{\overline{P}(x)}, \; x_1 = 1, 2, \ldots,$$

$P_{d,n}(x_n) = P(R_{d,,n} = x_n) = S_n(x_n)\, p(x_n)$,

where

$$S_n(x_n) = \sum_{0 \le x_0 < x_1 < \ldots < x_{n-1} < x_n} B(x_1)\, B(x_2) \ldots B(x_{n-1}),\ x_n = n-1, n, \ldots, \qquad (8.1.2)$$

The joint pmf of $R_{d,m}$ and $R_{d,n}$, $m < n$, is given by

$P_{d,m,n}(x_m, x_n) = P(R_{d,m} = x_m, R_{d,n} = x_n)$

$$= S_m(x_m)\, B(x_m)\, S_{m+1,n-1}(x_m, x_n)\, p(x_n),\ m \le x_m \le x_n - n + m < \infty, \qquad (8.1.3)$$

where

$$S_{m+1,n-1}(x_m, x_n) = \sum_{x_m < x_{m+1} < x_{m+2} < \ldots < x_n} B(x_{m+1}) \ldots B(x_{n-1}),\ m < n-1$$

$=1$, if $m = n-1$

The conditional pmf of $R_{d,n}$ given $R_{d,m} = x_m$ is given by

$P_{n \mid m}(x_n \mid R_{d,m} = x_m) = P(R_{d,n} = x_n \mid R_{d,m} = x_m)$

$$= S_{m,n}(x_m, x_n) \frac{p(x_n)}{\overline{P}(x_m)},\ x_m \le x_n - n + m < \infty. \qquad (8.1.4)$$

$$P_{n|n-1}(x_n \mid R_{d,n-1} = x_{n-1}) = \frac{p(x_n)}{\overline{P}(x_{n-1})},\ x_{n-1} < x_n. \qquad (8.1.5)$$

Using (8.1.1) and (8.1.4) it follows that the sequence of upper record values $R_{d(0)}$, $R_{d(1)}$, …, forms a Markov chain.

8.2. CHARACTERIZATIONS USING DISCRETE RECORDS

The following result was proved by Aliev and Ahsanullah (2002).

Theorem 8.2.1.

Let $X_1, X_2, \ldots$, be a sequence of independent and identically distributed random variables taking values on 0,1,2,..., with common distribution function F such that $F(n) < 1$ for all n and $E(X_i^2) < \infty$. Suppose that $\{B_k, k = 0,1,\ldots\}$ be a sequence of numbers such that $2 + 2B_{s+1} - B_s - B_{s+2} \geq 0$, s=0,1,2,.... If there exits F(x) such that $E\{I_{d(0,1)}^2 \mid R_{d,0} = s\} = B_s$, s =0, 1, 2..., where $I_{d(0,1)} = R_{d,1} - R_{d,0}$, then F(x) is unique.

Proof.

From (8.1.5), we obtain

$$E\{I_{d(0,1)}^2 \mid R_{d,0} = s\} = \frac{\sum_{j=1}^{\infty} j^2 p_{s+j}}{\sum_{j=1}^{\infty} p_{s+j}}, \; p_j = P(X = j), \; j = 0,1,2,\ldots, \tag{8.2.1}$$

Thus the condition $E\{I_{d(0,1)}^2 \mid R_{d,0} = s\} = B_s$ implies

$$\sum_{j=1}^{\infty} j^2 p_{s+j} = B_s \sum_{j=1}^{\infty} p_{s+j}, \text{ for } s \geq 0, \tag{8.2.2}$$

Writing s = s+1 in (8.1.2) , we get from (8.1.2)

$$\sum_{j=1}^{\infty} (j-1)^2 p_{s+j} = B_{s+1} \sum_{j=1}^{\infty} p_{s+j+1} \tag{8.2.3}$$

Subtracting (8.2.3) from (8.2.2) we obtain

$$\sum_{j=1}^{\infty} (2j-1) p_{s+j} = B_s \sum_{j=1}^{\infty} p_{s+j} - B_{s+1} \sum_{j=1}^{\infty} p_{s+1+j} \tag{8.2.4}$$

Now substituting s= s+1 in (8.2.4), we will have

$$\sum_{j=1}^{\infty} (2j-1) p_{s+1+j} = B_{s+1} \sum_{j=1}^{\infty} p_{s+1+j} - B_{s+2} \sum_{j=1}^{\infty} p_{s+2+j} \tag{8.2.5}$$

Subtracting (8.2.5) from (8.2.4) and on simplification, we obtain

$$p_{s+2} = \frac{1+2B_{s+1} - B_{s+2}}{B_{s+2}} p_{s+1} \quad - \frac{2+2B_{s+1} - B_s - B_{s+2}}{B_{s+2}} (1 - p_0 - p_1 - ...p_s) \quad (8.2.6)$$

Since the coefficients $-\frac{2+2B_{s+1} - B_s - B_{s+2}}{B_{s+2}}$ and $\frac{1+2B_{s+1} - B_{s+2}}{B_{s+2}}$ are positive, it means that p_{s+2} is increasing (decreasing) if p_{s+1} increases (decreases) for all $s \geq 0$. It means that for

any p=0, all probabilities p_2, p_3,....increases when p_1 increases. Together with the condition $\sum_{i=0}^{\infty} p_i = 1$, we conclude that for any given p_0, we have only one F(x) which will

satisfy (8.1.6).

8.3. Geometric Distribution

A discrete random variable X is said to have geometric distribution if its probability mass function(pmf) is of the following form:

$p(k) = P(X = k) = p\, q^{k-1}$, $0 < p < 1, q = 1\text{-}p, k \in A_0$

$= 0$, otherwise, (8.3.1)

where A_n = is the set of integers n+1, n+2, ..., and $n \geq 0$. We say $X \in GE(p)$, if the pmf of X is as given in (8.3.1). For $k > 0$, we define $r(k) = P[X = k | X \geq k]$.

We choose to distinguish between GE(p) and the larger class of distributions having geometric tail (GET). We write $X \in GET(s, p)$ if the pmf of X is as follows:

$p(k) = P[X = k] = cq^{k-1}$, $q = 1\text{-}p$, $k \in A_s$,

$= 0$, otherwise, (8.3.2)

where c is such that $\sum_{k=s+1}^{\infty} p(k) = 1$. If s = 0, then GET(s, p) = GE(p) with c = p.

The geometric distribution like the exponential distribution possesses the memory less property. That is;

$$\overline{P}(r+s) = \overline{P}(r)\overline{P}(s), \quad (8.3.3)$$

where r and s are positive integers and $\overline{P}(j) = \sum_{k=j+1}^{\infty} p(k)$. Geometric distribution is said to a discrete analogue of the exponential distribution.

8.3.1. Distribution of Record Values

If X ∈ GE(p), then

$$\overline{P}(x) = q^x$$

and

$p(x) = p\, q^{x-1}$, for $x \in A_0$.

Substituting the values of $\overline{P}(x_i)$ and $p(x_i)$ in (8.1.1), we get

$$P(R_{d,0} = x_0, R_{d,1} = x_1, \ldots, R_{d,n} = x_n) = \frac{p(x_0)}{\overline{P}(x_0)} \frac{p(x_1)}{\overline{P}(x_1)} \cdot \frac{p(x_2)}{\overline{P}(x_2)} \cdots \frac{p(x_{n-1})}{\overline{P}(x_{n-1})} p(x_n)$$

$$= p^{n+1} q^{x_n - n - 1}, \quad 1 \le x_0 < x_1 < x_2 < \ldots < x_n < \infty$$

$$= 0, \text{ otherwise.} \tag{8.3.4}$$

Let $V_0 = R_{d,0}$

$V_1 = R_{d,1} - R_{d,0}$

$V_n = R_{d,n} - R_{d,n-1}$

Then V_i's are independent and $V_i \in GE(p)$. We have

$R_{d,n} = V_0 + V_1 + V_2 + \ldots + V_n$.

If X ∈ GE(p), then

$$E(s^X) = \sum_{x=1}^{\infty} s^x p q^{x-1} = \frac{ps}{1 - qs}. \tag{8.3.5}$$

Using (8.3.5), we obtain

$$E(s^{R_{d,n}}) = E(s^{V_0+V_1+V_2+\dots V_n}) = \left(\frac{ps}{1-qs}\right)^{n+1} \quad (8.3.6)$$

The coefficient of s^x in $\left(\frac{ps}{1-qs}\right)^{n+1}$ is

$$\binom{x-1}{n} p^{n+1} q^{x-n-1}, x \geq n+1.$$

We can write

$$R_{d,n} | R_{d,m} = x \overset{d}{=} V_{m+1} + \dots + V_n + x \text{ , } n>m,$$

and

$$E(s^{R_{d,n}} \mid R_{d,m} = x) = s^x \left(\frac{ps}{1-qs}\right)^{n-m}.$$

The coefficient of s^y in $s^x \left(\frac{ps}{1-qs}\right)^{n-m}$ is

$$p_{n|m}(y \mid x) = P[R_{d,n} = y | R_{d,m} = x) = \binom{y-x-1}{n-m-1} p^{n-m} q^{y-x-n+m}.$$

Thus we obtain the conditional pmf of $R_{d,n}$ given $R_{d,m}$ as

$$P(R_{d,n} = x_n | R_{d,m} = x_m) = \binom{x_n - x_m - 1}{n-m-1} p^{n-m} q^{x_n - x_m - n + m} \text{ , } 0<m \leq x_m < x_n - n + m < \infty.$$

The joint pmf of $R_{d,m}$ and $R_{d,n}$ is

$$p_{m,n}(x,y) = P[R_{d,m} = x_m, R_{d,n} = x_n) = \binom{x-1}{m}\binom{y-x-1}{n-m-1} p^{n+1} q^{y-n-1},$$

$m+1 \leq x < y - n + m < \infty$

$= 0$, otherwise.

Let $I_{d(m,n)} = R_{d,n} - R_{d,m}$, $0 \leq m < n < \infty$, then

$$P(I_{d(m,n)} = z \mid R_{d,m} = x) = P(R_{d,n} = z + x \mid R_{d,m} = x)$$

$$= \binom{z-1}{n-m-1} p^{n-m}(1-p)^{z-n+m}, \text{ for } z \in A_{n-m-1},$$

$$= 0, \text{ otherwise.} \qquad (8.3.7)$$

Thus $I_{d(m,n)}$ and $R_{d,m}$ are independent. Further $I_{d(m,n)}$ and $R_{d,n-m}$ are identically distributed.

8.3.2. Characterizations of Geometric Distribution

There are several characterizations of geometric distributions based on (i) independence of $I_{d(m,n)}$ and $R_{d,m}$ (ii) conditional distribution of $R_{d,n}$ given $R_{d,m}$ for n >m and (iii) moment properties of some functions of $R_{d,n}$.

The following theorem (originally proved by Srivastava (1978)) is based on the probability of $I_{d(0,1)} \mid R_{d,0}$.

Theorem 8.3.1.

Suppose F(x) is the distribution function of the sequence of i.i.d. random variables $\{X_n, n \geq 1\}$ with positive mass function only at 1,2,... Then $P[I_{d(0,1)} = 1 \mid R_{d,0} = i] = P[I_{d(0,1)} = 1]$ for i = 1,2,..., if and only if X_n has the geometric distribution with pmf as given by

$$P_j = P[X = j] = c\,p\,(1-p)^{j-2},\ j = 2,3,.... \qquad (8.3.8)$$

and

$$p_1 = 1 - \sum_{j=2}^{\infty} p_j = 1 - c,\ 0 < p < 1,\ 0 < c < 1.$$

Proof.

We give here the original proof of Srivastasva (1979).

$$P[I_{d(0,1)} = 1 \mid R_{d,0} = i] = \frac{P[R_{d,1} = i+1, R_{d,0} = i]}{P[R_{d,0} = i]}$$

$$= \frac{p_{i+1}}{1-(p_1+...+p_i)}.$$

Since this conditional probability is independent of i, we must have

$$p = \frac{p_{i+1}}{1-(p_1+...+p_i)}, \text{ for } i = 1,2,...$$

For i = 1, we get $p_2 = p\,(1 - p_1\,) = c\ p$, say. For i = 2, we get

$$p_3 = p(1-p_1-p_2) = (1-p_1)p(1-p) = c\,p(1-p)\,.$$

Similarly for i = k, we have

$$p_k = c\,p(1-p)^{k-2}\,,\ k \geq 3. \qquad (8.3.9)$$

If p_j is as given by (8.3.9), then

$P[\ I_{d(0,1)} = 1 \mid R_{d,0} = i] = P\ [\ I_{d(0,1)} = 1]$.

A generalization of the Theorem 8.3.1 is the following theorem.

Theorem 8.3.2

Let $\{\ X_n,\ n \geq 1\}$ be a sequence of independent and identically distributed discrete random variables with common distribution function F. Suppose X is concentrated on the positive integers and $a = \sup\ \{\ x\ |F(x) < 1\} = \infty$. Then $X_n \in GET(n,p)$ for some fixed n, $n \geq 1$, if and only $I_{d(n,n+1)}$ and $R_{d,n}$ are independent.

Proof.

The 'only if ' part follows immediately from the from equation (8.3.5). So we need to established the ' if ' part. For $x \in A_0$, let

$c(u) = P[\ I_{d(n,n+1)} = u \mid R_{d,n} = x]$

$$= \frac{p(u+x)}{\overline{P}(x)}$$

$$= \frac{\overline{P}(u+x-1)-\overline{P}(u+x)}{\overline{P}(x)}, u \in A_0, x \in A_n. \quad (8.3.10)$$

Summing both sides of (8.3.10) with respect to u from 1 to u_0, and writing

$$c_1(u_o)=\sum_{u=1}^{u_o} c(u) \text{ and } c_o = 1- c_1(u_o) \quad . \quad (8.3.11)$$

on simplification, we get

$$\overline{P}(x+u_o)=c_o(u_o)\,\overline{P}(x),\ u \in A_o,\ x \in A_n. \quad (8.3.12)$$

The general solution of (8.1.12) is

$$\overline{P}(x)= cp^x,\ x \in A_n \quad (8.1.13)$$

where c is independent of p. Using the boundary condition $\overline{P}(\infty)= 0$, we get

$$\overline{P}(x)= cp^x,\ x \in A_n\ , 0 < p < 1, x \in A_n .$$

We have already seen that
$P[\ I_{d(n,n+1)} = u \mid R_{d,n}= y] = p\, q^{u-1}, u \in A_0,$

$= P[\ X_k = u].$

Does the above condition characterize the geometric distribution? As an answer to that question we have the following theorem.

Theorem 8.3.3.

Let $\{ X_n, n \geq 1\}$ be a sequence of independent and identically distributed random variables with common distribution function F. Suppose X is concentrated on the positive integers with a = sup $\{ x| F(x) < 1\} = \infty$. Further if $P[\ I_{d(n,n+1)} = u \mid R_{d,n}=y) = P[\ X_1 = u]$ for two fixed $y \in A_{n-1}$, where y_1, y_2 relatively prime to each other and all $u \in A_0$, then $X \in$ GET(n,p).

Proof.

Suppose that

$$P[\, I_{d(n,n+1)} = u \mid R_{d,n}=y) = P\,[\, X_1 = u], \tag{8.3.14}$$

then

$$P[\, I_{d(n,n+1)} = u \mid R_{d,n}=y)\,] = \frac{p(u+y)}{\overline{P}(y)} = p(u) \tag{8.3.15}$$

for two relatively prime $y_1, y_2 \in A_{n-1}$ and all $u \in A_0$. Summing (8.3.15) with respect to u from $u_0 + 1$ to ∞, we get

$$\frac{\overline{P}(u_o+y)}{\overline{P}(y)} = \overline{P}(u_o), \tag{8.3.16}$$

for two relatively prime $y_1, y_2 \in A_{n-1}$ and all $u \in A_0$. The general solution of (8.3.16) is

$$\overline{P}(x) = c\,p^x,\, x \in A_n,$$

and since $\overline{P}(\infty)=0$, we must have

$$\overline{P}(x) = c\,p^x, 0<p<1,\, x \in A_n. \tag{8.3.17}$$

Srivastava (1978) gave a characterization of the geometric distribution using the condition $E(R_{d(1)} \mid R_{d(0)} = y) = \alpha + y$. Ahsanullah and Holland (1984) proved the following theorem which is a generalization of Srivastava's result. The result is given in the following theorem.

Theorem 8.3.4.

Let $\{X_n, n \geq 1\}$ be a sequence of independent and identically distributed discrete random variables with common distribution function F. Suppose X is concentrated on the positive integers with $a = \sup\{x \mid F(x) < 1\} = \infty$. Further suppose $E[R_{d,n}]^2 < \infty$. If $E(R_{d,n+1} \mid R_{d,n} = y) = y + p^{-1}$ for all $y \in A_{n-1}$, then $X_1 \in GET(n,p)$ and $0 < p < 1$.

Proof.

Suppose

$$E(R_{d,n+1} \mid R_{d,n} = y] = p^{-1} + y, \text{ for all } y \in A_{n-1}.$$

It follows from (8.3.4) that

$$P[R_{d,n+1} = u+y \mid R_{d,n} = y] = \frac{p(u+y)}{\overline{P}(y)}, \; y \in A_{n-1},$$

and hence

$$E(R_{d,n+1} \mid R_{d,n} = y] =] = \sum_{u=y}^{\infty} (u+y)\frac{p(u+y)}{\overline{P}(y)}$$

$$= p^{-1} + y,$$

which implies that

$$\sum_{u=y}^{\infty} u \, \frac{p(u+y)}{\overline{P}(y)} = p^{-1}.$$

Thus we have

$$p^{-1} = \frac{\sum_{u=1}^{\infty} u[\overline{p}(u+y-1)-\overline{p}(u+y)}{\overline{P}(y)}$$

$$= \sum_{j=o}^{\infty} \frac{\overline{p}(y+i)}{\overline{P}(y)}$$

$$= \frac{J(y)}{J(y+1)-J(y+1)},$$

where

$$J(y) = \sum_{i=o}^{\infty} \overline{P}(y+i).$$

Thus

$$J(y+1) = q\, J(y), \; q = 1-p, \text{ for all } y \in A_{n-1}. \tag{8.3.18}$$

The general solution of (8.3.18) with the boundary condition, $J(n) = \theta$, $0 < \theta < 1$, is

$J(x) = \theta q^{x-n}$, $x \in A_{n-1}$.

Therefore

$\overline{P}(x) = J(x) - J(x+1) = \theta p q^{x-n}$, $x \in A_{n-1}$.

That is; $x \in GET(n,p)$.

8.4. Weak Records

Let $X_1, X_2, \ldots,$ be a sequence of independent identically distributed random variables taking values $0, 1, 2, \ldots, N$, $N \le \infty$ with a distribution function F such that $F(n) < 1$ for any n. Define the sequence of upper weak record times $T_{dw,n}$ as follows:

$T_{dw,()} = 1$,

$$T_{dw,n+1} = \min\{j > T_{dw,n}, X_j \ge X_{Tdw,n}\}, \; n = 0,1,2,\ldots, \tag{8.4.1}$$

The weak upper records are defined as

$$R_{wd,n} = X_{T_{dw,n}}, \; n = 0, 1, \ldots, \tag{8.4.2}$$

If we replace the sign $\ge$ by $>$ in (8.4.1) and (8.4.2), then we obtain upper record times and upper record values. Let

$$p_i = P(X_1 = i), \; q_i = p_i + p_{i+1} + \ldots + p_N.$$

Then the joint mass function of $R_{dw(0)}, R_{dw(1)}, \ldots, R_{dw(n)}$ is given by

$$P(R_{dw,0} = k_0, R_{dw,1} = k_1, \ldots, R_{dw,n} = k_n) = \prod_{i=0}^{n} \left(\frac{p_{k_i}}{q_{k_i}} \right) p_{k_n}, \tag{8.4.3}$$

see Vervaat(1973) for details. We now present the following characterization result.

Theorem 8.4.1.

Let $X_1, X_2, \ldots,$ be a sequence of independent identically distributed r.v.'s taking values $0,1,2,\ldots,N, N \leq \infty$ with a distribution function F such that $F(n) < 1$ for $n = 0,1,2,\ldots,$ and $E(X_1 \ln(1+X_1)) < \infty$. Let $R_{dw(n)}, n = 0,1,2,\ldots,$ be the upper weak record values from of the sequence {Xj, j=1,2,...}. Then the condition $E\left[\Psi(R_{dw,n+1}) \middle| R_{dw,n} = j\right) = g(j)$ determines the distribution.

Proof

It can be easily verified that

$$P\left(R_{dw,n+1} = y \middle| R_{dw,n} = x\right) = \frac{p_y}{q_x}, y \geq x.$$

and

$$E\left(\Psi(R_{dw,n+1}) \middle| R_{dw,n} = j\right] = \frac{1}{q_j} \sum_{k=j}^{N} \Psi(k) p_k. \tag{8.4.3}$$

If we take

$$E\left(\Psi(R_{dw,n+1}) \middle| R_{dw,n} = j\right] = g(j),$$

then equation (8.4.3) can be written as

$$g(j) q_j = \sum_{k=j}^{N} \Psi(k) p_k.$$

Taking the first order differences, we have

$$g(j) q_j - g(j+1)\, q_{j+1} = \Psi(j) p_j.$$

Thus

$$g(j)\, q_j - g(j+1)(q_j - p_j) = \Psi(j) p_j$$

or

$$p_j = \frac{g(j+1)-g(j)}{g(j+1)-\Psi(j)} q_j.$$

Since $q_j = \frac{q_j}{q_{j+1}} \cdot \frac{q_{j-1}}{q_{j-2}} \ldots \frac{q_1}{q_0}$, $q_0 = 1$, we have

$$p_j = \frac{g(j+1)-g(j)}{g(j+1)-\Psi(j)} \prod_{k=0}^{j-1} \left(\frac{q_{k+1}}{q_k} \right)$$

Also we obtain

$$g(j)q_j - g(j+1)q_{j+1} = \Psi(j)\left(q_j - q_{j+1}\right)$$

or

$$\frac{q_{j+1}}{q_j} = \frac{g(j)-\Psi(j)}{g(j+1)-\Psi(j)}.$$

Thus we have

$$p_j = \frac{g(j+1)-g(j)}{g(j+1)-\Psi(j)} \prod_{k=0}^{j-1} \left[\frac{g(k)-\Psi(k)}{g(k+1)-\Psi(k)} \right]$$

Hence the result follows.

Example 8.4.1.

Let $g(j) = j + b, N = \infty$ and $\Psi(j) = j$. In this case

$$p_j = \frac{1}{1+b}\left(\frac{b}{1+b}\right)^j, j = 0,1,2,\ldots,$$

i.e. X's are distributed as geometric. We now present the following extension

Theorem 8.4.2.

Let $X_1, X_2, \ldots,$ be a sequence of independent identically distributed r.v.'s taking values $0, 1, 2, \ldots, N$, $N \le \infty$ *with a distribution function* F *such that* $F(n) < 1$ *for* $n = 0, 1, 2, \ldots,$ *and* $E(X_1 \ln(1 + X_1)) < \infty$. *Let* $R_{dw,n}$ *be the sequence of weak records of the* X_i*'s . Then the condition* $E[\Psi(R_{dw,n+2} \mid R_{dw,n} = x\}] = ax + b$ *determines the distribution.*

Proof.

It can be verified that

$$P\left(R_{dw,n+2} = k \middle| R_{dw,n} = j\right) = \frac{p_k}{q_j} \sum_{r=j}^{k} \frac{p_r}{q_r},\ j \le k \le N \tag{8.4.4}$$

and

$$E[\Psi(R_{dw,n+2}) \mid R_{dw,n} = x] = \frac{1}{q_j} \sum_{r=j}^{k} \frac{p_r}{q_r} \sum_{k=r}^{N} \Psi(k) p_k \tag{8.4.5}$$

This along with the assumption gives

$$\frac{g(j)\, q_j^2}{p_j} - \frac{g(j+1) q_j\, q_{j+1}}{p_j} = \sum_{k=1}^{N} \Psi(k)\, p_k . \tag{8.4.6}$$

Taking the first order difference, we get

$$\frac{g(j) q_j^2 - g(j+1) q_j\, q_{j+1}}{p_j} - \frac{g(j+1)\, q_{j+1}^2 - g(j+2) q_{j+1}\, q_{j+2}}{p_{j+1}} = \Psi(j)\, p_j . \tag{8.4.7}$$

Dividing by q_j using $r(i) = q_i / q_{i+1} \left(r(i) - 1 = p_i / q_{i+1}\right)$, we obtain

$$\frac{g(j)\, r^2(j) - g(j+1) r(j)}{r(j) - 1} - \frac{g(j+1)\, r(j+1) - g(j+2)}{r(j+1) - 1} = \Psi(j) p_j . \tag{8.4.8}$$

Let $h(j) = 1/(r(j) - 1) = q_{j+1} / p_j$. We obtain on simplification, from (8.4.8)

$$h(j+1)=\frac{g(j)-g(j+1)}{g(j+1)-g(j+2)}(h(j)+2)+\frac{g(j)-\Psi(j)}{g(j+1)-g(j+2)}\frac{1}{h(j)}. \qquad (8.4.9)$$

The general solution of (8.4.9) for $h(j)$ is difficult. However, for selected values of $g(j)$ and $\Psi(j)$, solution of (8.4.9) for $h(j)$ can be obtained. For example if $g(j)=j+b$ and $\Psi(j)=j$, then $h(j)=b/2$.

Once a solution of equation (8.4.9) is obtained, p_j can be obtained as follows:

Let $h^*(j)$ be a solution of (8.4.9), then

$$p_j=\frac{1}{h^*(j)}\cdot\frac{q_{j+1}}{q_j}\cdot\frac{q_1}{q_{j-1}}\cdots\frac{q_1}{q_0},\ q_0=1.$$

Thus

$$p_j=\frac{1}{h^*(j)}\prod_{k=0}^{j-1}\frac{h^*(j)}{h^*(j)+1}. \qquad (8.4.10)$$

The following example illustrates the procedure.

Example 8.4.2.

If $g(j)=j+b, N=\infty$ *and* $\Psi(j)=j$, *then* $h^*(j)=b/2$ *and*

$$p_j=\frac{2}{b+2}\left(\frac{b}{b+2}\right)^j,\ j=0,1,\ldots,$$

i.e. X's are distributed as geometric.

Wesolowski and Ahsanullah (2001) proved the following theorem.

Theorem 8.4.3.

Let $X_1, X_2, \ldots,$ *be a sequence of independent identically distributed r.v.'s taking values* $0, 1, 2, \ldots, N$, $N\le\infty$ *with a distribution function F such that* $F(n)<1$ *for* $n=0,1,2,\ldots,$ *and* $E(X_1\ln(1+X_1))<\infty$. *Let* $R_{dw,n}$ *be the sequence of weak records of the* X_i*'s . Then the condition* $E[\Psi(R_{dw,n+2}|R_{dw,n}=x)]=ax+b$ for some real numbers a and b. Then a>0, b>0, and the following cases are possible:

(i) $0<a<1$, $b/(1-a)$ is a positive integer and

$$X \stackrel{d}{=} BB\left(\sqrt{a}/(1-\sqrt{a}), b/(1-a)\right);$$

(ii) $a=1$ and $X \stackrel{d}{=} GE(b/(2+b))$;

(iii) $a>1$ and $X \stackrel{d}{=} BNB\left((b+\sqrt{a}+1)/(a-1), b/(1-a)\right)$;

where the pmf's of the Beta-Binomial BB(β,N) and Beta-Negative Binomial BNB(β, γ) are respectively,

$$p_j = \binom{\beta+N-j-1}{N-J}\binom{\beta+N}{N}^{-1}, j=0,1,2,\ldots,N$$

and

$$p_j = \frac{\gamma}{\gamma+j}\binom{\beta}{\gamma}\binom{\beta+j+1}{\gamma+j}^{-1}, j=0,1,\ldots,$$

The proof can easily be obtained using Theorem 8.4.2.

REFERENCES

Ahsanullah, M. (1980). Linear Predictions of Record Values for two Parameter Exponential Distribution, *Ann. Inst. Statist. Math.* 32, 363-368.

Ahsanullah, M. (1995). *Introduction to Record Statistics*, NOVA Science Publishers Inc., Huntington, New York.

Ahsanullah, M. (2004). *Record Values- Theory and Applications.* University Press of America. Linham, MD,USA

Ahsanullah, M and Holland, B. (1984). Some Properties of Distributions of Record Values from Geometric Distribution. Statische Hefte 25, 319-317.

Ahsanullah, M. and Kirmani, S.N.U.A. (1991). Characterization of Exponential Distribution by Lower Record Values. Comm. Statist. Theory Meth. 20(4), 1293-1299.

Ahsanullah, M. and Malov, S. (2004). On ome characterizations via equi-distributional properties of records. J. Stat. Theory and Appl. (to appear).

Ahsanullah, M. and Nevzorov, V. B. (2001). *Ordered Random Variables,*NOVA Science Publishers Inc., Huntington, New York, USA.

Ahsanullah, M and Wesolowski, J. (1998). Linearity of Best Predictors for Non-adjacent Record Values. *Sankhyä,* Series B, 60, 221-227.

Aliev, F.A. and Ahsanullah, M. (2002). On characterizations of Discrete Distributions through Regressions of Record Values. *Pak. J. Stat.* 18,3, 415-421.

AlZaid, A.A. and Ahsanullah, M.(2003). A Characterization of the Gumbel Distribution Based on Record Values. *Comm. Stat. Theory and Meth.* 32,11, 2101-2108.

Arnold, B. C. (1985). P-Norm Bounds on the Expectation of the Maximum of Possibly Dependent Sample, *J. Multivariate Anal.* 17, 316-332.

Arnold, B. C, Balakrishnan, N. and Nagaraja, H.N. (1998). Records. New York, USA> *John Wile and Sons. Inc.*

Arnold, B.C. and Villasenor,J.A. (1997). Gumbel Records and Related Characterizations. In Johnson, N.L. and Balakrishnan,. Eds. Advances in the Theory and Practice of Statistics. A Volume in Honor of Samuel Kotz. New York, USA. *John Wiley and Sons.Inc*, 441-454.

Awad, A. M. and Raqab, M. Z. (2000). Prediction Intervals for the Future Record Values from Exponential Distribution: Comparative Study, *Journal of Statistical Computation and Simulation* 65(4), 325-340.

Arnold, B. C., Balakrishnan, N. and Nagaraja, H. N. (1998). *Records*, John Wiley, New York.

Awad, A. M. and Raqab, M. Z. (2000). Prediction Intervals for the Future Record Values from Exponential Distribution: Comparative Study, *Journal of Statistical Computation and Simulation* 65(4), 325-340.

Bairamov, I.G and M. Ahsanullah (2000). Distributional Relations Between Order Statistics and the Sample itself and Characterizations of Exponential Distribution. *Journal of Applied Statistical Science* 10(1), 1-16.

Basak, P.(1996).Lower Record Values and Characterizations of Exponential Distribution. *Cal. Stat. Asso, Bull.* 46,1-7.

Chandler, K. N. (1952). The Distribution and Frequency of Record Values, *J. Roy. Statist. Soc.*, Ser.B 14, 220-228.

Danielak, K. (2004). Distribution-Free Bounds for Expectations of Increments of Records, *Statistics and Probability Letters*, to appear.

Danielak, K. and Raqab, M.Z. (2003a): Sharp Bounds for Expectations of kth Record Increments, Australian and New Zealand Journal of Statistics, 69(4), to appear.

Danielak, K. and Raqab, M.Z. (2004b): Sharp Bounds on Expectations of kth Record spacings from Restricted Families, *Statistics and Probability Letters*, to appear.

Deminska, A. and Wesolowski, J. (2000).Linearity of Predictions for Non-adjacent Record Values. *J. Stat. Plan. And Inf.* 90,195-205.

Dunsmore, I. R. (1983). The Future Occurrenc of Records, *Ann.Inst. Statist. Math.* 35, 267-277.

Dziubdziela, W. and Kopoci'nski, B. (1976). Limiting Properties of the k-th Record Values, *Appl. Math.(Warsaw)* 15, 187-190.

Gajek, L. and Lenic, E. (1993). Moment Inequalities for Order and Record Statistics under Restrictions on their Distributions, *Ann. Univ. Mariae Curie-Skodowska*, Sec.A, 33, 27-35.

Gajek, L. and Okolewski, A. (1997). Steffensen-Type Inequalities for Order and Record Statistics, *Ann. Univ. Mariae Curie-Skodowska*, Sec. A, Vol. LI.1, 41-59.

Gajek, L. and Rychlik, T. (1996). Projection Method for Moment Bounds on Order Statistics from Restricted Families. II. Dependent Case, *J. Multivar. Anal.* 57, 156-174.

Gajek, L. and Rychlik, T. (1998). Projection Method for Moment Bounds on Order Statistics from Restricted Families. II. Independent Case, *J. Multivar. Anal.* 64, 156-182.

Gajek, L. and Okolewski, A. (2003). Projection Bounds on Expectations of Record Statistics from Restricted Families, *J. Statist. Plann. Infer.*, 110, 97-108.

Grudzie'n, Z. and Szynal, D. (1985). On the Expected Values of kth Record Values and Associated Characterizations of Distributions. In: Konecny, F., Mogyorody, J. and Wertz, W.,(eds.), *Probability and Statistical Decision Theory*, Vol. A, Reidel, Dordrecht, 119-127.

Gupta, R.C. and Ahsanullah, M. (2004). Some Characterization Results based on the Conditional Expectation of a function of non-adjacent order statistic (Record Value). Ann. Inst. Statist. Math.(to Appear).

Karlin, S. (1956). Decision Theory for Pólya Type Distribution, Case of Two Actions, I, Proceedings of the Third Berkeley Symposium on Statistics and Probability.

Karlin, S. and Studden, W. J. (1966). *Tchebyshev Systems: With Applications in Analysis and Statistics* , Wiley Interscience, New York.

Karlin, S.(1968). *Total Postivity*, Vol.I, Standard Univ. Press, Stanford, CA.

Kirmani, S.N.U.A. and Ahsanullah, M. (2001). Some results on Lower Records. Applied Statistical Science V. Ahsanullah, M., Kenyon, J. and Sarkar, S.K. eds. *Nova Science Publishers,Inc.* New York, USA, 77-85. v

Kirmani, S.N.U.A. and Beg, M.I. (1981) On Characterizations of Distributions by Expected Records. Sankhya 48 A, 463-465.

Lopez—Blazquez and Moreno Rebollo, J.L. (1007). A Characterization of Distributions Based on Linear Regression of Order Statistics and Record Values. *Sankhyä*, Series A, 59, 311-323.

Mitrinovic, D. S. (1972). *Analytic Inequalities*, Springer, Berlin.

Moriguti, S. (1953). A Modification of Schwarz's Inequality with Applications to Distributions, *Ann. Math. Statist.* 24, 107--113.

Nagaraja, H. N. (1977). On a characterization based on record values . *Austral. J. Statist.* 19,70-73.

Nagaraja, H. N. (1978). On the Expected Values of Record Values, *Austral. J. Statist.* 20, 176-182.

Nagaraja, H.N. (1990). Some Reliability Properties of Order Statistics, *Commun. Statist.-Theo.and Meth.* 19, 307-316.

Nevzorov, V.B.. (1988). Records. Theoretical Probability and Applications.32, 2, 201-228

Nevzorov, V.B.. (2001). Records. Mathematical Theory . Translations of Mathematical Monographs, Volume 194, Providence . RI, American Mathematical Society.

Rao, C.R. and Shanbag, D.N. (1994). Choquet-Deny Type Functional Equations and Applications to Stochastic Models. Wiley, New York.

Raqab, M. Z. (1997). Bounds Based on Greatest Convex Minorants for Moments of Record Values, *Statist. Probab. Lett.* 36, 35-41.

Raqab, M. Z. (2002). Characterizations of Distributions Based on the Conditional Expectations of Record Values. *Statist. And Decisions* 20, 200-315.

Raqab, M. Z. (2004a). Projection P-Norm Bounds on the Moments of k-th Record Increments, *J. Stat. Plan. .Inf.* 124(2), 301-315.

Raqab, M. Z. (2004b). Sharp Bounds On the Error in Approximating the Means of Record Statistics by Inverse Hazard Functions, *Commun. Statist. - Theory Meth.* 33(7), 1527-1539.

Raqab, M. Z. and Rychlik, T. (2002). Sharp Bounds for the Mean of the k-th Record Value, *Commun. Statist. - Theory Meth.* 31(11), 1927-1938.

Raqab, M. Z. and Rychlik, T. (2004). Sharp Bounds on the Expectations of Second Record Values from Symmetric Populations, *Statistics* 38(1), 49-57.

Resnick, S. I. (1973). Limit Laws for Record Values, *Stochastic Processes and Their Applications* 1, 67-82.

Rychlik, T. (1997). Evaluating Improvements of Records, *Appl. Math. (Warsaw)* 24, 315-324.

Rychlik, T. (2001). *Projecting Statistical Functionals*, Lectures Notes in Statistics Vol. 160, Springer-Verlag, New York.

Schoenberg. I. J. (1959). *On Variation Diminishing Approximation Mehtods. In: On Numberical Approximation*: Proceedings of a Symposium, Madison, WI, April 21-23, (R.E. Langer, Ed.), Publ. No. 1, Mathematics Research Center, USA Army, University of Wisconsin Press, Madison, WI.

Srivastava, R.C.(1978). Two Characterizations of the Geometric Distribution by Record Values. *Sankhyä*, Series B, 40, 276-278.

Takahasi, K. (1988). A Note on Hazard Rates of Order Statistics, *Commun. Statist. Theo. and Meth*. 17, 4133-4136.

Tata, M.N. (1969). On Outstanding Values in A Sequence of Random Variables, *Zeitschrift Fuer Wahrscheinlichkeitstheorie Und Verwandte Gebiete* 12, 9-20.

Vervat, W. Limit (1973).Theorems for Records for Discrete Distributions. Stoch. Proc. and Their Appl. 1, 317-334.

van Zwet, W.R. (1964). *Convex Transformations of Random Variables*, Mathematisch Centrum, Amsterdam.

Wesolowski , J. and Ahsanullah, M. (2001). Linearity of Regression for Non-adjacent Weak Records. Stat. Sinica, 11, 30-52.

Wu, J.W. (2004). On cCharacterizing Distributions by Conditional Expectations of Functions of Non-adjacent Record Values. *J. Appl. Statist. Sci*. (To appear).

INDEX

I

J

K

L

M

N

P

R

S

T

U

V

W

Y